Texts in Applied Mathematics

Volume 78

Editors-in-Chief

Anthony Bloch, University of Michigan, Ann Arbor, MI, USA

Charles L. Epstein, University of Pennsylvania, Philadelphia, PA, USA

Alain Goriely, University of Oxford, Oxford, UK

Leslie Greengard, New York University, New York, NY, USA

Series Editors

J. Bell, Lawrence Berkeley National Laboratory, Berkeley, CA, USA

R. Kohn, New York University, New York, NY, USA

P. Newton, University of Southern California, Los Angeles, CA, USA

C. Peskin, New York University, New York, NY, USA

R. Pego, Carnegie Mellon University, Pittsburgh, PA, USA

L. Ryzhik, Stanford University, Stanford, CA, USA

A. Singer, Princeton University, Princeton, NJ, USA

A. Stevens, University of Münster, Münster, Germany

A. Stuart, University of Warwick, Coventry, UK

T. Witelski, Duke University, Durham, NC, USA

S. Wright, University of Wisconsin, Madison, WI, USA

The mathematization of all sciences, the fading of traditional scientific boundaries, the impact of computer technology, the growing importance of computer modelling and the necessity of scientific planning all create the need both in education and research for books that are introductory to and abreast of these developments. The aim of this series is to provide such textbooks in applied mathematics for the student scientist. Books should be well illustrated and have clear exposition and sound pedagogy. Large number of examples and exercises at varying levels are recommended. TAM publishes textbooks suitable for advanced undergraduate and beginning graduate courses, and complements the Applied Mathematical Sciences (AMS) series, which focuses on advanced textbooks and research-level monographs.

Carl L. Gardner

Applied Numerical Methods for Partial Differential Equations

Carl L. Gardner
School of Mathematical & Statistical Sciences
Arizona State University
Tempe, AZ, USA

ISSN 0939-2475　　　　　　　　ISSN 2196-9949　(electronic)
Texts in Applied Mathematics
ISBN 978-3-031-69629-9　　　　ISBN 978-3-031-69630-5　(eBook)
https://doi.org/10.1007/978-3-031-69630-5

Mathematics Subject Classification: 65-01, 65F10, 65L04, 65M06, 65M08, 65M12, 76N15

© The Editor(s) (if applicable) and The Author(s), under exclusive license to Springer Nature Switzerland AG 2024

This work is subject to copyright. All rights are solely and exclusively licensed by the Publisher, whether the whole or part of the material is concerned, specifically the rights of translation, reprinting, reuse of illustrations, recitation, broadcasting, reproduction on microfilms or in any other physical way, and transmission or information storage and retrieval, electronic adaptation, computer software, or by similar or dissimilar methodology now known or hereafter developed.
The use of general descriptive names, registered names, trademarks, service marks, etc. in this publication does not imply, even in the absence of a specific statement, that such names are exempt from the relevant protective laws and regulations and therefore free for general use.
The publisher, the authors and the editors are safe to assume that the advice and information in this book are believed to be true and accurate at the date of publication. Neither the publisher nor the authors or the editors give a warranty, expressed or implied, with respect to the material contained herein or for any errors or omissions that may have been made. The publisher remains neutral with regard to jurisdictional claims in published maps and institutional affiliations.

This Springer imprint is published by the registered company Springer Nature Switzerland AG
The registered company address is: Gewerbestrasse 11, 6330 Cham, Switzerland

If disposing of this product, please recycle the paper.

Frontispiece

Cylindrically symmetric simulation by C. L. Gardner, J. R. Jones, E. Scannapieco, and R. A. Windhorst [15] of star formation by the Centaurus A jet emitted by the active galactic nucleus of NGC 5128, using a parallelized WENO3 method. The jet is propagating to the right at Mach 1000 with respect to the ambient gas, and star clusters form inside the bow-shocked molecular clouds (*orange/red circles*). The logarithm of density is shown at 1 Myr. The computational domain is 6300 light years on each side. More physically realistic 3D simulations are presented in Fig. 8.21.

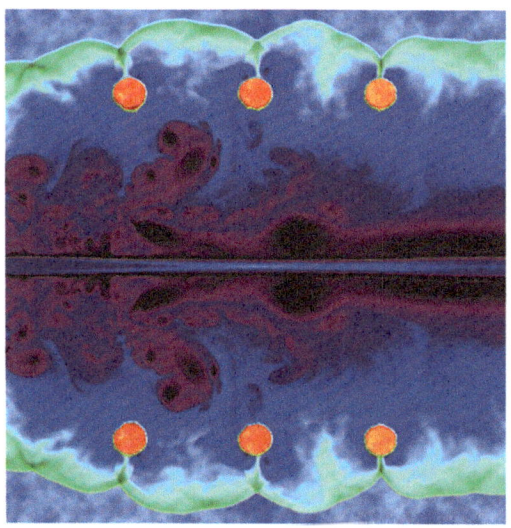

For Mark DeLamotte Gardner, Paul DeLamotte Gardner, and Denise Tessier Scarduzio

Preface

The aim of this book is to rapidly advance students to the level where they can solve systems of linear and nonlinear partial differential equations (PDEs) with state-of-the-art finite difference and finite volume numerical methods: TRBDF2 (trapezoidal rule/backward difference formula second-order) for parabolic modes like heat conduction or diffusion, modern iterative methods including PCG (preconditioned conjugate gradients) and GMRES (generalized minimal residuals) for elliptic modes like electrostatics or potential flow in fluids, and WENO (weighted essentially non-oscillatory [method]) for hyperbolic modes like advection or gas dynamics. We also review state-of-the-art numerical methods for systems of linear and nonlinear ordinary differential equations (ODEs) including the explicit fourth-order Runge-Kutta method and the implicit TRBDF2 method. (Abbreviations for various mathematical methods and concepts are explained in the Acronyms section on p. xvii.)

The book can serve as the text for a one-semester graduate course on the numerical solution of partial differential equations, or—omitting some of the more advanced material like gas dynamics and PDEs of mixed type—for a one-semester junior/senior or graduate course on the numerical solution of ordinary and partial differential equations.

Topics not usually treated in introductory courses on the numerical solution of ODEs and PDEs include solving (i) chaotic dynamical systems with TRBDF2, (ii) nonlinear boundary value problems with Newton's method, (iii) the nonlinear diffusion equation with TRBDF2 and Newton's method, (iv) the nonlinear Laplace equation with Newton's method and GMRES, (v) the inviscid Burgers equation with WENO3 (third-order WENO), (vi) gas dynamics with WENO3 (1D Riemann problems and 2D supersonic jets), (vii) the drift-diffusion model with TRBDF2, and (viii) the classical hydrodynamic model (electro-gas dynamics) with WENO3 plus TRBDF2.

In particular, most introductions to numerical methods for PDEs give short shrift to hyperbolic PDEs, often covering only the classical methods (first-order upwind, Lax-Friedrichs, and Lax-Wendroff) for the advection and wave equations, and completely omitting not only gas dynamics (the paradigmatic example of a nonlinear system of hyperbolic conservation laws) and the best methods for nonlinear hyperbolic PDEs—higher-order upwind methods like WENO, PPM (piecewise parabolic method), or CLAWPACK (Conservation Laws Package)—but even omitting the inviscid Burgers equation and any mention of nonlinear hyperbolic waves. We give a full introduction to gas dynamics, as well as Burgers' equation, and their nonlinear waves, plus WENO3 solvers for these equations.

While TRBDF2 is the method of choice for linear and nonlinear parabolic PDEs, there are a number of other excellent methods like multigrid for elliptic PDEs and higher-order Godunov methods like PPM or CLAWPACK for hyperbolic PDEs. Our goal is to provide students with one or two of the best methods in each case, for the sake of speed of exposition.

Consistency, *stability*, and *convergence* of numerical methods are stressed.

Newton's method is used throughout to linearize nonlinear ODE boundary value problems and nonlinear elliptic PDEs, as well as in implicit methods like backward Euler, TR, and TRBDF2 for time-dependent nonlinear ODEs and PDEs.

Many of the time-dependent PDEs in mathematical physics—and science in general—can be expressed as conservation laws. *Conservative* numerical methods are *always* advocated for PDE conservation laws.

Model programs illustrating each of the numerical methods are provided in MATLAB and are easily translatable[1] to other modern programming languages like C, C++, Fortran, Python, etc. The 34 programs are available at

> https://github.com/carl-parsec/codes4book

Parallel methods including chaotic relaxation and domain decomposition are discussed at a number of points, but are not included in the model programs.

The 93 problems are an integral aspect of the book, confirming quoted results in the text and exploring new directions. In addition to the problems at the end of each chapter, eight semester-long projects are also suggested there.

Major scientific applications include:

- Lorenz equations
- Boundary layer equation
- Heat/diffusion equation
- Nonlinear diffusion (semiconductor process simulation)
- Laplace's and Poisson's equations (electrostatics)
- Nonlinear Laplace equation
- First-order wave (linear advection) equations
- Second-order wave equation
- Burgers' equation
- Gas dynamics (1D Riemann problems and 2D supersonic jets)
- Drift-diffusion model (ion flow in biological channels and cells)
- Navier-Stokes equations
- Classical hydrodynamic model (semiconductor device simulation)

---

[1] For large-scale scientific computing, C++ or the most recent version of Fortran should be used.

By the end of the book, students will be able to implement state-of-the-art numerical methods for systems of linear and nonlinear PDEs of mixed type, involving a combination of parabolic, elliptic, and/or hyperbolic behavior.

My sincere thanks are due to my longtime colleague and collaborator Jeremiah Jones (School of Mathematical & Statistical Sciences, Arizona State University), for the joint development of the MATLAB WENO3 code (based on a Fortran implementation of Guang-Shan Jiang and Chi-Wang Shu), and for the simulations of the Centaurus A astrophysical jet (Frontispiece plus Fig. 8.21) and the triad synapse of the retina (Fig. 9.1).

Tempe, AZ, USA
July 2024

Carl L. Gardner

Contents

1 **Overview** 1
   1.1   Code Validation 2
   1.2   Recommended Numerical Methods 3
          1.2.1   Recommended Methods for ODEs 3
          1.2.2   Recommended Methods for PDEs 5
   1.3   Code Development 6

2 **Consistency, Stability, and Convergence** 9
   2.1   Derivative Approximations 9
   2.2   IEEE Floating Point 11
   2.3   Numerical Errors 13
   2.4   No Units on Digital Computers! 13
   2.5   Equivalence Theorem 15
   2.6   A-Stability and L-Stability 17
   2.7   Stability for PDEs 18
   2.8   Local Truncation Error vs. Local Error 18
   Problems 20

3 **Numerical Methods for ODE IVPs** 21
   3.1   Forward and Backward Euler Methods 22
   3.2   Trapezoidal Rule Method 24
   3.3   Predictor-Corrector Methods 25
   3.4   Runge-Kutta Methods 26
   3.5   Harmonic Oscillator 27
   3.6   Dynamic Timestep 28
   3.7   Deterministic Dynamical Systems 32
          3.7.1   Nonlinear Pendulum 32
          3.7.2   Van der Pol Oscillator 34
          3.7.3   Shaw Oscillator 34
          3.7.4   Lorenz Equations 36
          3.7.5   Stability of Equilibria 37
   3.8   Backward Euler and Newton's Method 39

	3.9	TRBDF2 Method	40
	3.10	Equivalence Theorem for ODE IVPs	43
	Problems	45
4	**Numerical Methods for ODE BVPs**	**49**
	4.1	First and Second Derivative Matrices	49
	4.2	Linear Boundary Value Problems	50
	4.3	Nonlinear BVPs and Newton's Method	51
	Problems	55
5	**Overview of PDEs**	**57**
	5.1	Classifying Linear PDEs	58
	5.2	Classifying General Nonlinear Systems of PDEs	60
	5.3	Equivalence Theorem for PDE IVPs	62
	Problems	63
6	**Numerical Methods for Parabolic PDEs**	**65**
	6.1	Heat/Diffusion Equation	68
	6.2	Fourier Solution to the Diffusion Equation	69
	6.3	Conservative Form	71
	6.4	Parabolic Conservation Laws	73
	6.5	Forward and Backward Euler Methods	73
	6.6	Implicit vs. Explicit Methods for the Diffusion Equation	75
	6.7	Von Neumann Stability Analysis	75
	6.8	Trapezoidal Rule Method	76
	6.9	Relationship of Parabolic PDEs to ODE IVPs	76
	6.10	Boundary Conditions for Parabolic/Elliptic PDEs	77
		6.10.1 Dirichlet Boundary Conditions	77
		6.10.2 Neumann Boundary Conditions	77
	6.11	Nonlinear Diffusion: TRBDF2 and Newton's Method	78
		6.11.1 TRBDF2 Revisited	78
		6.11.2 Dynamic Timestep	80
		6.11.3 Accuracy and Stability	80
		6.11.4 Jacobian $\partial f / \partial u$	81
		6.11.5 TRBDF2 Simulations of Nonlinear Diffusion	82
	6.12	Finite Volume Method for Multidimensional PDEs	83
	Problems	86
7	**Numerical Methods for Elliptic PDEs**	**91**
	7.1	Laplace's and Poisson's Equations	93
	7.2	Fourier Solution for the Model Laplace Problem	96
	7.3	The Classical Iterative Idea	97
	7.4	Classical Iterative Methods for Laplace's Equation	99
		7.4.1 Jacobi Iteration	100
		7.4.2 Gauss-Seidel Iteration	101
		7.4.3 SOR Iteration	101
	7.5	Theory of Classical Iterative Methods	104

Contents

	7.6	Conjugate Gradients	107
		7.6.1 Method of Steepest Descent	108
		7.6.2 Conjugate Gradient Method	108
		7.6.3 CG Algorithm	110
		7.6.4 PCG Algorithm	110
	7.7	GMRES	111
		7.7.1 GMRES Algorithm	112
		7.7.2 Arnoldi Iteration Algorithm	112
	7.8	Nonlinear Laplace Equation: Newton's Method	113
		7.8.1 Discretization and Newton's Method	113
		7.8.2 Simulations of the Nonlinear Laplace Equation	116
	Problems		117
8	**Numerical Methods for Hyperbolic PDEs**		121
	8.1	First-Order Wave (Linear Advection) Equations	124
	8.2	Second-Order Wave Equation	126
	8.3	Transformation to Characteristic Variables	127
	8.4	First-Order Upwind Method	128
	8.5	Modified Equation and Stability	129
	8.6	Von Neumann Stability Analysis	131
	8.7	Lax-Friedrichs Method	132
	8.8	Hyperbolic Conservation Laws	133
	8.9	Lax-Wendroff Method	134
	8.10	Two-Step Lax-Wendroff Method	135
	8.11	Rankine-Hugoniot Jump Conditions	137
	8.12	Solutions to Nonlinear Hyperbolic Conservation Laws	139
	8.13	Riemann Problem for Nonlinear Hyperbolic Conservation Laws	140
	8.14	Inviscid Burgers Equation and Riemann Problem	140
	8.15	Numerical Methods for the Inviscid Burgers Equation	145
	8.16	Ideal Gas Dynamics	149
	8.17	Gas Dynamical Shocks and Contacts	151
	8.18	Characteristic Equations for Gas Dynamics	153
	8.19	Riemann Problem for Gas Dynamics	155
	8.20	Boundary Conditions for Gas Dynamics	156
		8.20.1 Periodic Boundary Conditions	157
		8.20.2 Through-Flow Boundary Conditions	157
		8.20.3 Wall Boundary Conditions	158
	8.21	Space-Time Finite Volume Methods for Conservation Laws	159
	8.22	Dimension by Dimension Splitting	160
	8.23	WENO Method	160
	8.24	WENO3 Simulations of Riemann Problems	166
	8.25	2D WENO3 Simulations of Supersonic Jets	167
	8.26	WENO3 Codes	170
	8.27	Godunov's Method	172
	8.28	Loss of Accuracy at a Shock Wave	175
	Problems		177

9 Numerical Methods for Mixed Type PDEs ... 185
- 9.1 Strang Splitting ... 185
- 9.2 Compressible Navier-Stokes Equations ... 187
- 9.3 Drift-Diffusion Model ... 188
- 9.4 Incompressible Navier-Stokes Equations ... 191
- 9.5 Classical Hydrodynamic Model ... 194
- Problems ... 196

A Useful Mathematical Formulas ... 201
- A.1 MATLAB vs. C Indices ... 201
- A.2 Taylor Series Formulas ... 201
  - A.2.1 Taylor Series for $f(x)$ ... 201
  - A.2.2 Taylor Series for $u(t)$ ... 202
  - A.2.3 Taylor Series for $f(x, y)$ ... 202
  - A.2.4 Taylor Series for $u(x, t)$ ... 202
- A.3 Basic Finite Difference Derivatives ... 202
  - A.3.1 Spatial Derivatives ... 202
  - A.3.2 Forward Time Difference ... 203
- A.4 Orthogonality Relations for Fourier Series ... 203

B Norms and Condition Number ... 205
- B.1 l_p Vector Norms for a Vector of Fixed Length ... 205
- B.2 l_p Vector Norms for a Vector of Variable Length n ... 205
- B.3 Basic Properties of Vector Norms ... 206
- B.4 Matrix Norms ... 206
- B.5 Condition Number ... 207

References ... 209

Index ... 211

Acronyms

BC	boundary condition
BVP	boundary value problem
CFL	Courant-Friedrichs-Lewy
CG	conjugate gradient
CLAWPACK	Conservation Laws Package (by LeVeque)
ENO	essentially non-oscillatory (*pronounced* E-NO)
FDTD	finite-difference time-domain
FFT	fast Fourier transform
flop	floating point operation [a combined add-multiply $a*b+c$ or a divide a/b]
GMRES	generalized minimal residuals (*pronounced* G-M-REZ)
IVP	initial value problem
KdV	Korteweg-De Vries
LF	Lax-Friedrichs
LTE	local truncation error
LW	Lax-Wendroff
MMR	modified midpoint rule
ODE	ordinary differential equation
PCG	preconditioned conjugate gradients
PDE	partial differential equation
PPM	piecewise parabolic method
RK	Runge-Kutta
RP	Riemann problem
SOR	successive over-relaxation
SUR	successive under-relaxation
TR	trapezoidal rule
TRBDF2	trapezoidal rule/backward difference formula second-order
TVD	total variation diminishing
WENO	weighted essentially non-oscillatory (*pronounced* WE-NO)

Notation

$h \equiv \Delta x$
Wave number $k = 2\pi/\lambda$, where λ is the wavelength.

Spatial and Temporal Partial Derivatives

For a function $u = u(x, t)$,

$$u_t = \frac{\partial u}{\partial t}, \quad u_x = \frac{\partial u}{\partial x}$$

$$u_{tt} = \frac{\partial^2 u}{\partial t^2}, \quad u_{xx} = \frac{\partial^2 u}{\partial x^2}, \quad u_{xt} = \frac{\partial^2 u}{\partial t \partial x}, \quad u_{tx} = \frac{\partial^2 u}{\partial x \partial t}$$

$$\partial_t = \frac{\partial}{\partial t}, \quad \partial_x = \frac{\partial}{\partial x}$$

Spatial and Temporal Indices

For time-dependent PDEs, the exact solution $u(x_i, t_n) \approx u_i^n$.
For time-dependent ODEs, $u(t_n) \approx u_n$.
For ODE boundary value problems, $y(x_i) \approx y_i$.
For elliptic PDEs, $u(x_i, y_j, z_k) \approx u_{ijk}$.

Kronecker δ_{ij}

$$\delta_{ij} = \begin{cases} 1 \text{ if } i = j \\ 0 \text{ if } i \neq j \end{cases}$$

Derivative Matrices

$$\text{tridiag}[a, b, c] \equiv \begin{bmatrix} b & c & 0 & 0 & 0 \\ a & b & c & 0 & 0 \\ 0 & \ddots & \ddots & \ddots & 0 \\ 0 & 0 & a & b & c \\ 0 & 0 & 0 & a & b \end{bmatrix} \equiv \begin{bmatrix} b & c & & & \\ a & b & c & & \\ & \ddots & \ddots & \ddots & \\ & & a & b & c \\ & & & a & b \end{bmatrix}$$

White space in matrices denotes 0s.
Central first derivative matrix

$$\mathcal{D}^{(1)} = \frac{1}{2\Delta x} \text{tridiag}[-1, 0, 1]$$

Central second derivative matrix

$$\mathcal{D}^{(2)} = \frac{1}{\Delta x^2} \text{tridiag}[1, -2, 1]$$

Backward first derivative matrix

$$\mathcal{D}^{(-)} = \frac{1}{\Delta x} \text{tridiag}[-1, 1, 0]$$

Forward first derivative matrix

$$\mathcal{D}^{(+)} = \frac{1}{\Delta x} \text{tridiag}[0, -1, 1]$$

Computer Codes

Our computer programs (all in MATLAB) are available at

> https://github.com/carl-parsec/codes4book

The programs are all self-contained for simplicity of use. For example, **chd.m** has a built-in WENO3 solver identical to the one in **weno3.m**.

burgers1.m	conservative upwind method and four versions of the Lax-Friedrichs method for the inviscid Burgers equation
burgers3.m	WENO3 method for the inviscid Burgers equation
bvp.m	solves a linear BVP using central differences and a tridiagonal direct solve
chd.m	WENO3 method for the classical hydrodynamic model for semiconductor devices
diffusion0.m	backward Euler method for the linear diffusion equation with fixed timestep and Dirichlet BCs
diffusion1.m	TRBDF2 method for the linear diffusion equation with fixed timestep and Dirichlet BCs
diffusion2.m	TRBDF2 method for the linear diffusion equation with dynamic timestep and Dirichlet BCs
diffusionNBC.m	TRBDF2 method for the linear diffusion equation with dynamic timestep and homogeneous Neumann BCs
diffusion2D.m	TRBDF2 method for the 2D linear diffusion equation with fixed timestep for movie and Dirichlet BCs
ivp.m	solves the IVP $dy/dt = -y$, $y(0) = y_0$, using forward Euler, backward Euler, and TR
jets.m	simulates 2D gas dynamical supersonic jets using the WENO3 method
laplace0.m	banded matrix direct solve for the 2D Laplace equation with Dirichlet BCs
laplace1.m	Jacobi iteration for the 2D Laplace equation with Dirichlet BCs

laplace2.m	*Gauss-Seidel iteration for the 2D Laplace equation with Dirichlet BCs*
laplace3.m	*SOR iteration for the 2D Laplace equation with Dirichlet BCs*
laplace4.m	*conjugate gradient method for the 2D Laplace equation with Dirichlet BCs*
laplace5.m	*PCG method for the 2D Laplace equation with Dirichlet BCs*
laplace6.m	*MATLAB built-in CG, PCG, or GMRES method for the 2D Laplace equation with Dirichlet BCs*
laplaceNBC.m	*SOR method for the 2D Laplace equation with a Neumann BC on one side*
layer.m	*solves the nonlinear layer BVP using central differences by Newton iteration (with a tridiagonal direct solve)*
lorenz1.m	*solves the Lorenz equations using fourth-/fifth-order Runge-Kutta or TRBDF2*
lorenz2.m	*solves the Lorenz equations with two different sets of initial conditions using fourth-/fifth-order Runge-Kutta*
lorenz3.m	*solves the Lorenz equations with three different methods or sets of initial conditions using fourth-/fifth-order Runge-Kutta*
lorenzeigs.m	*calculates the eigenvalues of the Lorenz Jacobian at the equilibria of the Lorenz equations*
nonlin_diffusion.m	*solves the nonlinear diffusion equation with homogeneous Neumann BCs using TRBDF2 with dynamic timestep and Newton's method (with a tridiagonal direct solve)*
nonlin_laplace.m	*solves the 2D nonlinear Laplace equation with Dirichlet BCs using Newton's method with a banded matrix direct solve or GMRES*
pendulum.m	*plots the phase space diagram/direction field for the nonlinear pendulum*
rk2.m	*solves the IVP $dw/dt = f(w)$ using second-order Runge-Kutta with fixed timestep*
rk2dyn.m	*solves the IVP $dw/dt = f(w)$ using second-order Runge-Kutta with dynamic timestep*
shaw.m	*solves the Shaw equations using fourth-/fifth-order Runge-Kutta*
wave1.m	*solves the first-order wave equation using the original upwind method*
wave2.m	*solves the second-order wave equation as a first-order system using the Lax-Friedrichs or Lax-Wendroff method*
wave3.m	*solves the first-order wave equation using the WENO3 method*
weno3.m	*solves 1D gas dynamical Riemann problems using the WENO3 method with Lax-Friedrichs flux splitting*

Chapter 1
Overview

Differential equations describe how quantities change in time and/or space, or more generally how quantities change with respect to other quantities. The fundamental laws of physics, as well as many theoretical and applied problems in mathematics, the sciences, engineering, medicine, and economics are formulated in terms of ordinary and partial differential equations (abbreviated as ODEs and PDEs).

Linear ODEs and PDEs can often be solved exactly, while analytical solutions to nonlinear ODEs and PDEs are few and far between. Those that are known are justly famous: Newton's planetary orbits,[1] the Riemann problem solution in gas dynamics, Taylor vortex solutions of the incompressible Navier-Stokes equations, the Cole-Hopf solution of Burgers' equation, solitons in the Korteweg-De Vries equation, or the Schwarzschild and Friedmann solutions of Einstein's equations of general relativity, to mention the author's favorites. Thus, nonlinear ODEs and PDEs generally are solvable only through numerical simulation.

There are three differential equation classes with related numerical methods: (i) ODE initial value problems and parabolic PDEs, (ii) ODE boundary value problems and elliptic PDEs, and (iii) hyperbolic PDEs. The first and third classes involve evolution in time, while the second class is time-independent. We will review numerical methods for ODE initial and boundary value problems first and then develop numerical methods for parabolic, elliptic, and hyperbolic PDEs in turn. *Consistency*, *stability*, and *convergence* of the numerical methods are analyzed throughout.

We begin with an overview of code validation—how do we know whether a numerical simulation is correct or not?—and a summary of recommended numerical methods for ODEs and PDEs. *The reader may wish to return to this short overview throughout the course as the numerical concepts and methods are developed.*

---

[1] From solving Newton's second law with gravity, yielding a nonlinear ODE for the planetary position $\mathbf{r}(t)$: $d^2\mathbf{r}/dt^2 = -GM\,\mathbf{r}/|\mathbf{r}|^3$.

© The Author(s), under exclusive license to Springer Nature Switzerland AG 2024
C. L. Gardner, *Applied Numerical Methods for Partial Differential Equations*,
Texts in Applied Mathematics 78, https://doi.org/10.1007/978-3-031-69630-5_1

Modern time-dependent methods are written as (possibly composite) one-step methods, because one-step methods are easy to start and restart, and memory usage is most efficient. Thus, all of our recommended time-dependent numerical methods are one-step.

Each of the methods mentioned below is discussed later, with model programs in MATLAB, except we omit the fast Fourier transform (FFT) method for the Poisson/Laplace equation (see [35], e.g.), multigrid methods (see [4]), adaptive mesh refinement (see [3]), and the Finite-Difference Time-Domain (FDTD) method (see [37]) for Maxwell's equations as subjects for other books and courses. Godunov's method is presented, but the reader is referred to the PPM (piecewise parabolic method) [9] and CLAWPACK (Conservation Laws Package) [25] references and websites for documentation and code. Additionally, although we discuss Chorin's method [6] for the Navier-Stokes equations, a Navier-Stokes solver is relegated to the suggested projects.

For the sake of speed of exposition, we concentrate on TRBDF2 (trapezoidal rule/backward difference formula second-order) [1] and fourth-order Runge-Kutta (RK4) for ODEs, TRBDF2 for parabolic PDEs like the diffusion equation, modern iterative methods including PCG (preconditioned conjugate gradients) [20] and GMRES (generalized minimal residuals) [31] for elliptic PDEs like the Laplace and Poisson equations, and WENO (weighted essentially non-oscillatory [method]) [33] for hyperbolic PDEs like the wave equations and gas dynamics.

Many of the time-dependent PDEs in mathematical physics—and science in general—can be expressed as conservation laws, including the heat/diffusion equation, the wave equations, Maxwell's equations, Burgers' equation, compressible fluid dynamics, the drift-diffusion model, the classical hydrodynamic model, general relativity, etc. *Conservative* numerical methods are *always* recommended for PDE conservation laws.

Generally speaking, second-order accuracy is sufficient for almost all scientific problems, although higher-order methods like PPM (third-order) and WENO5 (fifth-order in space, third-order in time) are required for the sharpest resolution of strong shock waves in gas dynamics. However, even for the case of strong shock waves, second-order CLAWPACK is competitive with PPM and WENO5. There are also ODE initial value problems like orbital mechanics that require higher-order methods for sufficient accuracy (e.g., launching a spacecraft from Earth to orbit Mars).

Note that we always use dimensionless units for digital computations, since we are computing just with dimensionless numbers (zeros and ones on modern binary computers): *"There are no units on digital computers!"*

1.1 Code Validation

An important scientific question is "How do we know if simulation results are correct?" There are several strategies:

- Demonstrate convergence under spatial mesh refinement $\Delta x \to 0$ (in 3D, $\Delta x \to 0$, $\Delta y \to 0$, and $\Delta z \to 0$), and/or temporal mesh refinement $\Delta t \to 0$. (See the 2D spatial grid in Fig. 2.1.) Figure 1.1 shows 2D simulations of a supersonic astrophysical jet with radiative cooling emitted by a young star on three successively finer grids. Note the associated bow shock and the Kelvin-Helmholtz rollup of the tip of the jet. The large-scale coarse grid features are reproduced on the medium and fine grids, while additional small-scale features appear on the finer grids, and a 2000×2000 simulation shows the same structural details as the 1000×1000 simulation—this is convergence under mesh refinement. Also note that since $\Delta t \sim h$ for hyperbolic PDEs, where $h \equiv \Delta x = \Delta y$, we are refining the timestep as we refine the spatial mesh.
- Compare with *science* and experiment. The astrophysical jet simulations can be compared with telescopic images of astrophysical jets.
- Compare with exact or approximate analytical solutions: for example, the approximately linear regime of a nonlinear problem or the Riemann problem for 1D gas dynamics (Fig. 1.2).
- Vary physical parameters: For example, in Burgers' equation, vary viscosity for a shock profile solution (see Fig. 8.9).
- Compare with results using other numerical methods and other programs: For example, for time-dependent ODEs, try an explicit solver like fourth-order Runge-Kutta vs. an implicit solver like TRBDF2 (see Problem 3.13); for the inviscid Burgers equation, compare upwind vs. Lax-Friedrichs vs. WENO3 (see Sect. 8.15 and Problem 8.14); in gas dynamics, try WENO3 vs. CLAWPACK (see [19]).
- Check symmetry and conservation. The symmetry about the horizontal $y = 0.5$ jet axis is not imposed in Fig. 1.1, so the observed reflection symmetry serves as a check on the numerical method used in the simulations.

1.2 Recommended Numerical Methods

Use Newton's method to linearize nonlinear ODE boundary value problems and nonlinear elliptic PDEs, as well as in implicit methods like TRBDF2 for time-dependent nonlinear ODEs and PDEs.

1.2.1 Recommended Methods for ODEs

For ODE initial value problems, always dynamically adjust Δt based on an estimate of the local error.

Fig. 1.1 $250\Delta x \times 250\Delta y$, $500\Delta x \times 500\Delta y$, and $1000\Delta x \times 1000\Delta y$ 2D Mach 80 astrophysical jet simulations with radiative cooling using WENO3 (third-order in space and time): $\log_{10}(n/\bar{n})$ with $\bar{n} = 100$ H atoms/cm^3. The grid is 10^{11} km on each side

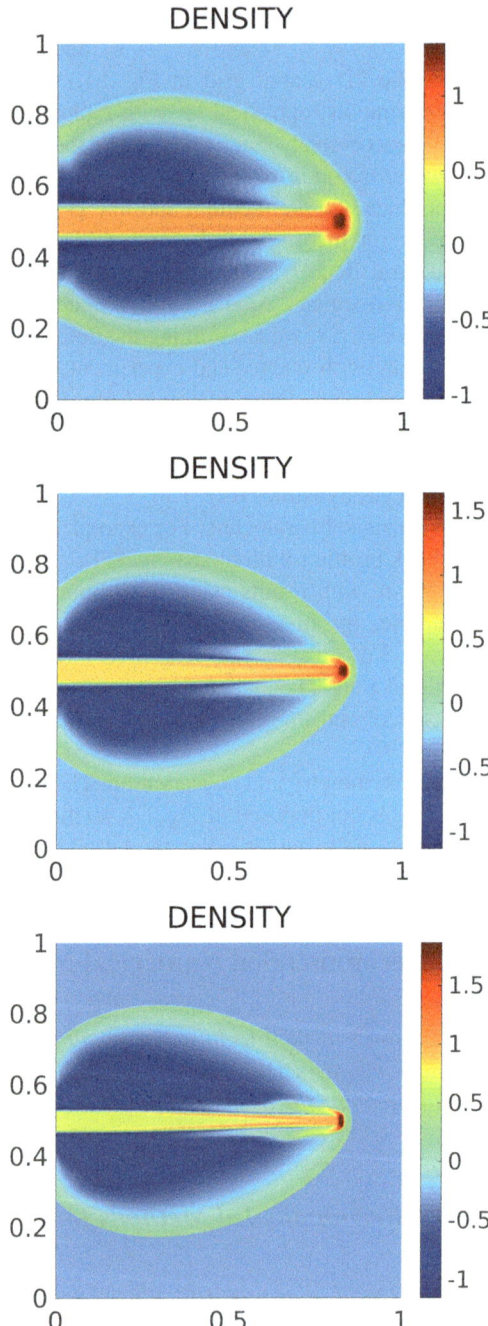

1.2 Recommended Numerical Methods

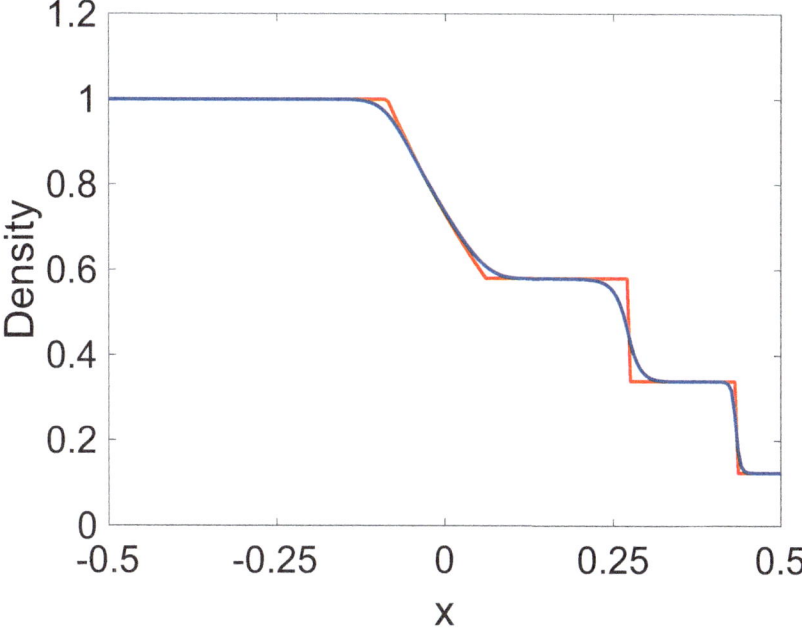

Fig. 1.2 WENO3 (*smoother curve*) simulation of density computed using **weno3.m** with $200\Delta x$ at $t = 0.2$ vs. exact solution of a Riemann problem with (from the left) rarefaction wave, contact wave, and shock wave. Even better spatial resolution can be obtained with $400\Delta x$ (see Fig. 8.23) or WENO5

- TRBDF2 for ODE initial value problems, especially for *stiff* problems (which have two or more widely disparate solution timescales), or fourth-order Runge-Kutta.
- Newton's method for nonlinear ODE boundary value problems.

1.2.2 Recommended Methods for PDEs

For PDE conservation laws, always use a conservative numerical method: Discretize $(D(u)u_x)_x$ in 1D or $\nabla \cdot (D(u)\nabla u)$ in 3D directly; always use conserved quantities as the solution variables. Adaptive mesh refinement can be used in spatial regions where more resolution is required.

For parabolic PDEs, always dynamically adjust Δt based on an estimate of the local error. For hyperbolic PDEs, Δt is adjusted dynamically based on the CFL (Courant-Friedrichs-Lewy) condition for stability: $\Delta t \leq \Delta x / \max\{|\lambda|\}$, where $\max\{|\lambda|\}$ is the maximum characteristic speed in magnitude. For some hyperbolic schemes, there may be a fractional constant in the CFL condition, for example, $\Delta t \leq \frac{1}{2}\Delta x / \max\{|\lambda|\}$.

- *Parabolic PDEs:* TRBDF2 for linear and nonlinear diffusion or drift-diffusion; Chorin's method plus predictor-corrector timestep for Navier-Stokes; TR (trapezoidal rule) for Schrödinger's equation.
- *Hyperbolic PDEs:* Conservative upwind for wave equations and the inviscid Burgers equation; WENO or higher-order Godunov (PPM or CLAWPACK) for gas dynamics; FDTD (related to Lax-Wendroff) for Maxwell's equations.
- *Elliptic PDEs:* For the Laplace and Poisson equations, banded or sparse matrix direct solvers, or Chebyshev SOR (successive over-relaxation), PCG, GMRES, or multigrid iterative solvers. Use Newton's method to linearize nonlinear problems.

1.3 Code Development

The best approach to developing a computer code for a nonlinear differential equation consists of the following steps, illustrated by the example of writing a third-order WENO code for 3D gas dynamics:

- First develop an upwind code for the linear wave equation

$$u_t + cu_x = 0. \tag{1.1}$$

Compare computed and exact $u(x,t) = u_0(x - ct)$ solutions.
- Then extend the upwind code to the nonlinear inviscid Burgers equation

$$u_t + \left(\frac{1}{2}u^2\right)_x = 0. \tag{1.2}$$

Here we have exact shock wave, rarefaction wave, and N-wave solutions. In each case, compare the computed solution with the exact solution.
- Next solve the inviscid Burgers equation with the WENO3 method, again comparing the computed and exact solutions.
- Use the WENO3 method to solve 1D gas dynamics

$$w_t + f(w)_x = 0 \tag{1.3}$$

applying the code to Riemann problems and then comparing computed and exact solutions for paradigmatic Riemann problems (there are eight implemented in **weno3.m**).
- Extend the **weno3.m** code to handle 2D (see **jets.m**) and 3D gas dynamical problems, using direction by direction splitting (two or three copies of the 1D solver). Now we are generally solving problems where we do not have an exact or even approximate analytical solution, but we can compare computed solutions against experiment, theory, and numerical simulations with other codes.

1.3 Code Development

In 2D, we can compute solutions to the 2D Riemann problems given in [27] and compare with computed solutions presented there obtained with eight different state-of-the-art numerical methods.
- Now we are ready to solve 2D and 3D research problems, perhaps with additional physical effects like radiative cooling, etc.

Chapter 2
Consistency, Stability, and Convergence

The three central themes of this book are consistency (accuracy), stability, and convergence. The three concepts are interrelated by the Lax-Richtmyer Equivalence Theorem—also known as the Fundamental Theorem of Numerical Analysis.

Theorem *Lax-Richtmyer Equivalence Theorem: For* consistent *numerical approximations,* stability *and* convergence *are equivalent.*

We will develop both intuitive and mathematically precise definitions of these three concepts. Three proofs of the theorem will be presented: (i) due to Lax and (ii) due to Strang in Sect. 3.10 for ODE initial value problems; and (iii) due to Richtmyer in Sect. 5.3 for PDE initial value problems.

Derivative approximations provide a simple example of discretization error, while an introduction to IEEE floating point motivates the study of roundoff error and stability.

2.1 Derivative Approximations

Finite difference approximations of spatial and temporal derivatives are developed using Taylor series expansions (reviewed in Appendix A.2).

In 1D space with $x \in [a, b]$, we discretize on a spatial grid (as along the x axis in Fig. 2.1) with fixed steps $\Delta x = (b-a)/N$ and grid points

$$x_i = a + i\Delta x, \quad i = 0, 1, \ldots, N. \tag{2.1}$$

Note that $x_0 = a$ and $x_N = b$. (In MATLAB, we label grid points $x_1 = a$, $x_2, \ldots$, x_N, $x_{N+1} = b$. See Appendix A.1.) Discretization in 3D adds spatial grid points y_j and z_k. Given a function $f(x)$, grid point values are defined by $f_i = f(x_i)$; for $f(\mathbf{x})$, grid point values are defined by $f_{ijk} = f(x_i, y_j, z_k)$.

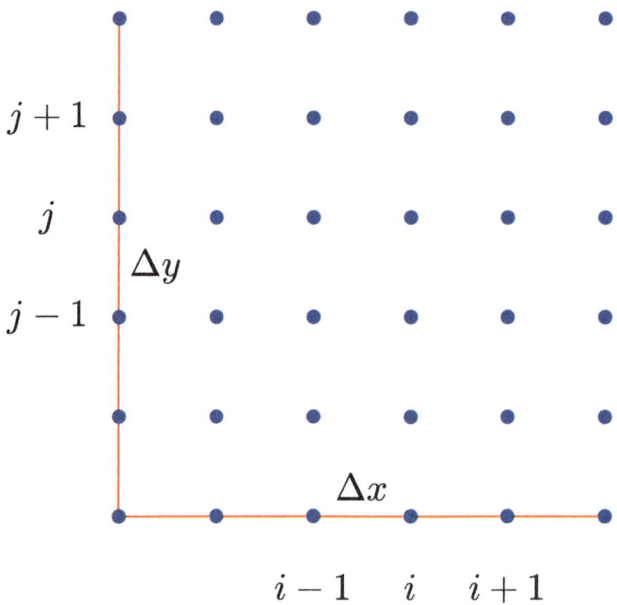

Fig. 2.1 2D spatial grid points

For time-dependent problems, we discretize on a temporal grid with variable timesteps Δt_n and temporal points

$$t_{n+1} = t_n + \Delta t_n, \quad n = 0, 1, 2, \ldots \quad (2.2)$$

Given a function $u(t)$, time level values are defined by $u_n = u(t_n)$.

As an example, the upwind first derivative for flow to the right is

$$\left(\frac{df}{dx}\right)_i \approx \frac{f_i - f_{i-1}}{\Delta x} = \frac{f(x_i) - f(x_{i-1})}{\Delta x} = \frac{f(x_i) - f(x_i - \Delta x)}{\Delta x}$$

$$= \frac{1}{\Delta x}\left(f_i - \left(f_i - \Delta x f_i' + \frac{\Delta x^2}{2}f_i'' - \cdots\right)\right) = f_i' - \frac{\Delta x}{2}f_i'' + \cdots$$

$$(2.3)$$

This approximation is consistent and first-order accurate since the leading term as $\Delta x \to 0$ on the right-hand side of (2.3) is correctly $(df/dx)_i$ and the leading error term is proportional to Δx.

Below we list the first-order and second-order accurate first and second derivatives (these formulas are also recorded in Appendix A.3). Verification of the formulas is assigned to the problems at the end of the chapter.

Second-order accurate central difference approximation to first derivative:

2.2 IEEE Floating Point

$$\left(\frac{df}{dx}\right)_i \approx \frac{f_{i+1} - f_{i-1}}{2\Delta x} = f'_i + \frac{\Delta x^2}{6} f'''_i + \cdots \quad (2.4)$$

First-order accurate backward difference approximation to first derivative:

$$\left(\frac{df}{dx}\right)_i \approx \frac{f_i - f_{i-1}}{\Delta x} = f'_i - \frac{\Delta x}{2} f''_i + \cdots \quad (2.5)$$

First-order accurate forward difference approximation to first derivative:

$$\left(\frac{df}{dx}\right)_i \approx \frac{f_{i+1} - f_i}{\Delta x} = f'_i + \frac{\Delta x}{2} f''_i + \cdots \quad (2.6)$$

Second-order accurate central difference approximation to second derivative:

$$\left(\frac{d^2 f}{dx^2}\right)_i \approx \frac{f_{i+1} - 2f_i + f_{i-1}}{\Delta x^2} = f''_i + \frac{\Delta x^2}{12} f_i^{(4)} + \cdots \quad (2.7)$$

Note that the central difference second derivative at i is given by differencing the central difference first derivatives at $i \pm \frac{1}{2}$:

$$f''_i \approx \frac{f'_{i+\frac{1}{2}} - f'_{i-\frac{1}{2}}}{\Delta x} \approx \frac{f_{i+1} - 2f_i + f_{i-1}}{\Delta x^2} \quad (2.8)$$

with $f'_{i+\frac{1}{2}} = (f_{i+1} - f_i)/\Delta x$ and $f'_{i-\frac{1}{2}} = (f_i - f_{i-1})/\Delta x$.

Forward time difference:

$$\frac{du}{dt} \approx \frac{u_{n+1} - u_n}{\Delta t} = \frac{u(t_{n+1}) - u(t_n)}{\Delta t} = \frac{u(t_n + \Delta t) - u(t_n)}{\Delta t} = u' + \frac{u''}{2}\Delta t + \cdots \quad (2.9)$$

With this formula, the forward and backward Euler methods for time-dependent differential equations are first-order accurate in time, while the TR method is second-order accurate in time.

2.2 IEEE Floating Point

Moler's *Numerical Computing with MATLAB* [29], Sect. 1.7, gives an excellent summary of IEEE floating point, on which this section is based.

Computer arithmetic is performed in digital computation according to the IEEE floating point standard (1985). To understand sources of error in digital computation, we need to understand floating point arithmetic.

We will always use double precision floating point numbers with 64 bits—single precision is not sufficient for scientific computing, and double precision floating point is almost as fast as single precision on contemporary computers.

In IEEE floating point, a nonzero real number $\bar{r}$ is represented on the computer by

$$\bar{r} \approx r = \pm(1+f)\,2^n \tag{2.10}$$

where the 1 is a *phantom* (i.e., not stored, but built into the computer arithmetic). In double precision floating point, there are 64 bits for r: One bit is allotted for the sign s ($s = 0$ for a $+$ sign and $s = 1$ for a $-$ sign), the mantissa $0 \leq f < 1$ is allotted 52 bits (precision), and the exponent $-1022 \leq n \leq 1023$ is allotted the remaining 11 bits (range).

The number r is stored as $\{s, f, x\}$, where the shifted exponent

$$1 \leq x = n + 1023 \leq 2046 = 2^{11} - 2. \tag{2.11}$$

$x = 0$ is reserved for *underflow*, with 0 represented exactly by all zeros in $\{s, f, x\}$, and -0 by all zeros except $s = 1$; $x = 2^{11} - 1$ corresponds to Inf (infinity) if $f = 0$ and to NaN (not a number) if $f \neq 0$.

In MATLAB, check that: $1/0 = \text{Inf}$, $1/\text{Inf} = 0$, $\text{Inf} + \text{Inf} = \text{Inf}$, $0/0 = \text{NaN}$, $\text{Inf} - \text{Inf} = \text{NaN}$.

Note that $\frac{1}{10}$ cannot be represented exactly in finite precision binary floating point, since

$$\frac{1}{10} \text{ (base 10)} = 0.1\overline{1001} \text{ (base 2)}. \tag{2.12}$$

Try $0.3/0.1 - 3$ in MATLAB. The answer is not 0 on the computer, because the floating point numerator in $0.3/0.1$ is a little less than 0.3 and the floating point denominator in $0.3/0.1$ is a little greater than 0.1.

Machine epsilon ϵ_M is defined, so that $1 + \epsilon_M$ is the next positive number after 1 on the computer. Verify that

$$\epsilon_M = 2^{-52} \approx 2.22 \times 10^{-16} \tag{2.13}$$

$$realmax = (2 - \epsilon_M)\,2^{1023} \approx 2^{1024} \approx 1.8 \times 10^{308} \tag{2.14}$$

$$realmin = 2^{-1022} \approx 2.2 \times 10^{-308}. \tag{2.15}$$

2.4 No Units on Digital Computers!

Either ϵ_M or $\epsilon_M/2$ may be called the roundoff level. Try[1] $(1 + 0.51 * \text{eps}) - 1$ and $(1 + 0.5 * \text{eps}) - 1$ in MATLAB. There are 53 significant bits or 16 significant digits base 10 in IEEE double precision, but only 24 significant bits or 7 significant digits base 10 in IEEE single precision (32 bits = 1 sign bit + 23 bit mantissa + 8 bit exponent). MORAL: *Always use double precision.*

There *may* also be a range of *subnormal* numbers down to $\epsilon_M \, realmin$ on the computer. Try eps * realmin in MATLAB.

2.3 Numerical Errors

There are three types of numerical error that are intrinsic to digital computing:

(i) *Roundoff error:* Floating point operations introduce an additional fractional error of ϵ_M. In N floating point operations[2] (*flops*), roundoff errors can accumulate randomly (error $\sim \sqrt{N}\epsilon_M$ as in a random walk), systematically (error $\sim N\epsilon_M$), or catastrophically ($1/(b-a)$ can $\to \pm\text{Inf}$). IEEE floating point arithmetic uses *round half to even*. Then roundoff errors grow $\sim\sqrt{N}$ rather than linearly $\sim N$.

(ii) *Truncation or discretization error:* An infinite series or sequence has to be approximated by a finite number of terms in floating point arithmetic. For example, a continuous real interval $[a, b]$ is approximated by a finite set of grid points $x_0 = a, x_1 = x_0 + \Delta x, x_2 = x_0 + 2\Delta x, \ldots, x_N = x_0 + N\Delta x = b$; or, as another example,

$$e^x \approx \sum_{n=0}^{N} \frac{x^n}{n!}, \quad |x| < 1. \tag{2.16}$$

(iii) *Numerical instability:* Roundoff error introduced early in the computation can be exponentially amplified to swamp the true solution. For example, the forward Euler method for the heat/diffusion equation $u_t = u_{xx}$ is stable only if $\Delta t \leq \Delta x^2/2$ (Fig. 2.2).

2.4 No Units on Digital Computers!

Since we are computing just with dimensionless numbers (in fact zeros and ones), we are always using dimensionless units for digital computations. Of course

---

[1] In MATLAB, ϵ_M is represented by eps.
[2] A combined add-multiply or a divide.

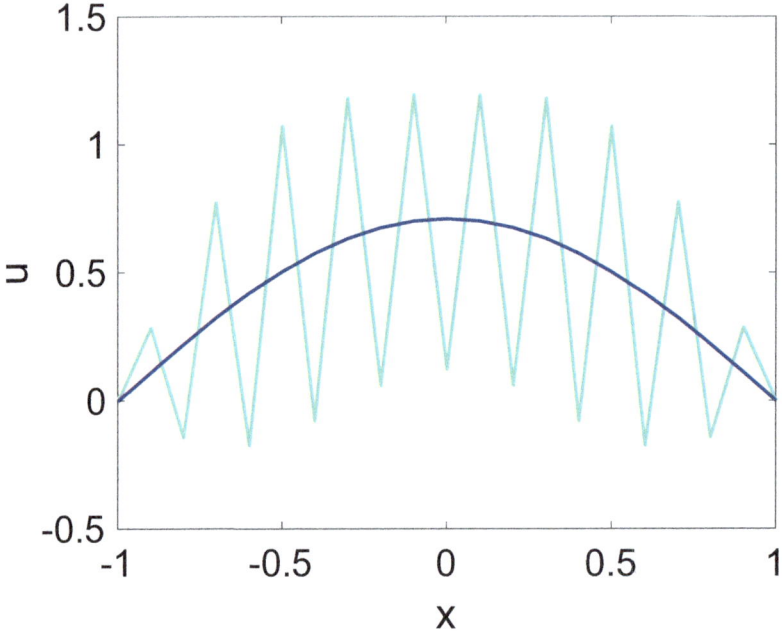

Fig. 2.2 Stable (*smooth*, $\Delta t = \Delta t_{FE}$) and unstable (*jagged*, $\Delta t = 1.2\,\Delta t_{FE}$) forward Euler computed solutions to the heat/diffusion equation $u_t = u_{xx}$ at $t = 0.18$ with $20\Delta x$ (to see the oscillations in the unstable solution clearly), where $\Delta t_{FE} = \Delta x^2/2$ is the maximum stable Δt for forward Euler

computed dimensionless physical quantities like a dimensionless velocity $\tilde{v} = v_{phys}/\bar{v}$ can be converted to the correct physical units (say to m/s if $\bar{v} = r$ m/s, where r is a positive number) on output by multiplying $v_{output} = r\,\tilde{v}$. The scale factor r is chosen so that the magnitudes of typical computed velocities are $O(1)$ in order to minimize the effects of roundoff error during computation.

An example will suffice to make this issue clear. Consider (for simplicity) 1D linear diffusion during semiconductor processing:

$$\frac{\partial u}{\partial t} = D\frac{\partial^2 u}{\partial x^2}, \quad u(x, t=0) = u_0(x) \qquad (2.17)$$

where $u(x, t)$ is the number density of diffusing particles (dopant) and $D > 0$ is the diffusion coefficient. Typical values for boron diffusing in Si are $u \sim 1.5 \times 10^{20}$ boron atoms/cm^3 in a 0.25 micron gate device for approximately 1000 s. The physical diffusion coefficient is $D = 2 \times 10^{-13}$ cm^2/s. Beginning students are tempted to use units like cm, s, atoms/cm^3, and cm^2/s for length, time, number density, and the diffusion coefficient, respectively, as computational units (or SI units). Instead though of working with numbers like 10^{20}, 10^{-4}, 10^3, and 10^{-13} (or

10^{26}, 10^{-6}, 10^3, and 10^{-17}), let's choose computational scales[3]

$$\bar{x} = 10^{-5}\,\text{cm}, \quad \bar{t} = 10^3\,\text{s}, \quad \bar{u} = 10^{20}\,\text{cm}^{-3}. \tag{2.18}$$

Then the derived computational scale $\bar{D} = \bar{x}^2/\bar{t} = 10^{-13}\,\text{cm}^2/\text{s}$. Now our typical computational values are $\tilde{u} = u/\bar{u} \sim 1.5$ in an $\tilde{x} = x/\bar{x} = 2.5$ gate device for approximately $\tilde{t} = t/\bar{t} = 1$. The computational diffusion coefficient is $\tilde{D} = D/\bar{D} = 2$.

We scale the diffusion Equation (2.17) by multiplying by $\bar{t}/\bar{u}$

$$\frac{\partial(u/\bar{u})}{\partial(t/\bar{t})} = (D/\bar{D})\frac{\partial^2(u/\bar{u})}{\partial(x/\bar{x})^2}, \quad u(x/\bar{x}, t/\bar{t} = 0)/\bar{u} = u_0(x/\bar{x})/\bar{u} \tag{2.19}$$

or in dimensionless computational variables

$$\frac{\partial \tilde{u}}{\partial \tilde{t}} = \tilde{D}\frac{\partial^2 \tilde{u}}{\partial \tilde{x}^2}, \quad \tilde{u}(\tilde{x}, \tilde{t} = 0) = \tilde{u}_0(\tilde{x}). \tag{2.20}$$

In programming the diffusion equation solution, we drop the tildes and always use computational units.

2.5 Equivalence Theorem

The three types of numerical error are intimately related to the ideas of consistency, stability, and convergence. We can now make these concepts in the Equivalence Theorem (*stability* and *convergence* are equivalent for *consistent* numerical methods) more precise. Proofs of the Equivalence Theorem are delayed until the appropriate material is covered: Sect. 3.10 for ODE initial and boundary value problems and Sect. 5.3 for PDE initial value problems.

The basic idea of *stability* of a numerical method for an initial value problem is that roundoff error, which is inextricably introduced at each timestep, is *not* exponentially amplified during the course of a computation. More precise details are provided immediately below, after we define consistency and convergence.

A numerical method is *consistent* and *p*th-order accurate for the ODE initial value problem $du/dt = f(u)$, where u and f are in general vectors, if the leading local error is proportional to Δt^{p+1}, $p > 0$. Assuming stability, the leading (global) error then will be proportional to Δt^p. Typically p is an integer, though it might be $\frac{1}{2}$, etc.

---

[3] Because the diffusion equation is homogeneous of degree one in u, we can independently choose units for u as well as for x and t.

A numerical method is *convergent* if the numerical solution u_n goes to the exact solution $u(t_n)$ in the limit in which $\Delta t \to 0$ and $n \to \infty$, with $t_n = n\Delta t$ held fixed:

$$||u(t_n) - u_n|| \to 0. \tag{2.21}$$

See Appendix B for the definitions of popular vector norms. If u is a scalar, $||u|| = |u|$.

For linear initial value problems, the numerical solution at time level n is given by $u_n = G^n u_0$, where G is the growth factor of the numerical method. The method is *stable* if for $du/dt = au$, $\mathfrak{R}\{a\} < 0$, $||G|| \leq 1$; or equivalently, if numerical errors introduced by roundoff in floating point arithmetic are *not* exponentially amplified.

To analyze the stability of a numerical method for the nonlinear initial value problem $du/dt = f(u)$ (where for PDEs f is an operator functional), we linearize the initial value problem with respect to a small perturbation δu about a given nonlinear solution $\bar{u}$. Setting $u(t) = \bar{u}(t) + \delta u(t)$,

$$\frac{d\bar{u}}{dt} + \frac{d\delta u}{dt} = f(\bar{u} + \delta u) \approx f(\bar{u}) + J\,\delta u \tag{2.22}$$

$$\frac{d\delta u}{dt} \approx J\,\delta u, \quad J = \left.\frac{\partial f}{\partial u}\right|_{\bar{u}}. \tag{2.23}$$

Then we analyze the stability of the numerical method for the linearized initial value problem.

A-stability and L-stability are defined for ODE initial value problems in Sect. 2.6, and for PDE initial value problems in Sect. 2.7 based on von Neumann's Fourier analysis of stability. There are more specific stability analyses (Lax-Richtmyer stability, method of lines stability, etc.), which the interested reader is encouraged to investigate (see the discussions in [26] and references therein).

Strictly speaking, von Neumann's Fourier analysis of stability applies only to linear PDEs with constant coefficients, and only for the Cauchy problem (no boundary conditions). The analysis can be extended to the case of periodic boundary conditions, by considering a Cauchy problem with periodic initial data.

Stability with more general boundary conditions can be difficult to analyze. With nonconstant coefficients, the coefficients can be frozen to analyze stability, and for nonlinear PDEs, linearization about a nonlinear solution can be used. Our PDE methods, which are stable under von Neumann's Fourier analysis, are always stable in these more general contexts.

There are some stability proofs for nonlinear PDEs, but they require functional analysis and apply only to a few specialized cases.

2.6 A-Stability and L-Stability

A time integration method for $du/dt = au$, $\Re\{a\} < 0$, is *A-stable* or *absolutely stable* if

$$||G|| \leq 1 \tag{2.24}$$

for all $\Delta t > 0$. The time integration method is *L-stable* if it is A-stable and

$$\lim_{\Delta t \to \infty} ||G|| = 0. \tag{2.25}$$

For ODE initial value problems and the heat/diffusion equation, the TR method is A-stable but not L-stable, while the backward Euler (first- and second-order) and TRBDF2 methods are L-stable.

The difference between an A-stable method like TR (which may have local oscillations for large Δt) vs. an L-stable method like TRBDF2 is illustrated in Fig. 2.3.

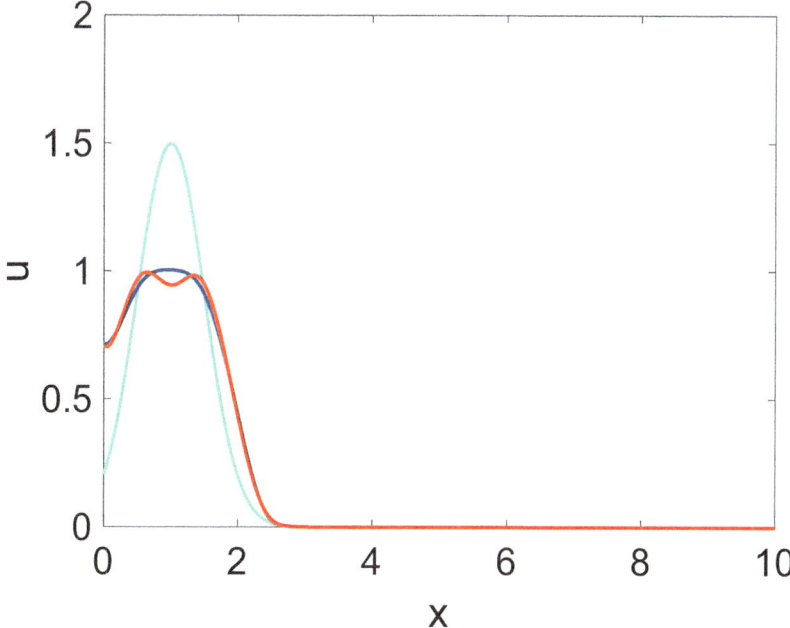

Fig. 2.3 Simulation of nonlinear diffusion of dopant concentration in semiconductor processing using **nonlin_diffusion.m** on a 1 micron domain with $200\Delta x$ and homogeneous Neumann boundary conditions, with one large timestep $\Delta t = 0.1$. The initial conditions $u(x, t = 0)$ are plotted (high peak). The local oscillation (or "ringing") in the A-stable (but not L-stable) TR solution is spurious, while the L-stable TRBDF2 method gives good results

2.7 Stability for PDEs

To analyze stability for linear time-dependent PDEs, we will use von Neumann's Fourier analysis method. General initial data at time t can be represented by a Fourier integral

$$u(x, t) = \int_{-\infty}^{\infty} \frac{dk}{2\pi} e^{ikx} u_k(t) \tag{2.26}$$

where $k = 2\pi/\lambda$ is the wave number (recall $e^{ikx} = \cos(kx) + i\sin(kx)$) and λ is the wavelength of the Fourier mode. Then after a timestep of Δt,

$$u(x, t + \Delta t) = \int_{-\infty}^{\infty} \frac{dk}{2\pi} e^{ikx} u_k(t + \Delta t) \equiv \int_{-\infty}^{\infty} \frac{dk}{2\pi} e^{ikx} G(k) u_k(t). \tag{2.27}$$

Now set the initial data u_j^n at time level n and grid point j equal to a single Fourier mode $u_j^n = e^{ikx_j}$, and derive the growth factor $u_j^{n+1} = G(k) u_j^n$ for this mode. Then for stability, we require $||G(k)|| \leq 1$ for all k.

The Fourier analysis of stability for linear PDEs can be extended to the linearized version of nonlinear PDEs.

2.8 Local Truncation Error vs. Local Error

In solving the initial value problem

$$\frac{du}{dt} = f(u), \quad u(t=0) = u_0 \tag{2.28}$$

with a consistent numerical method, we make a local error[4] $e_l^m = c_m \Delta t^{p+1}$, $p > 0$, in going from time level $m - 1$ to m, where the c_m are constants. Heuristically, the (global) error after n timesteps of Δt with $t_n = n\Delta t$ fixed, assuming stability,[5] is approximately

$$e \sim \sum_{m=1}^{n} c_m \Delta t^{p+1} \sim n \bar{c} \Delta t^{p+1} = t_n \bar{c} \Delta t^p \tag{2.29}$$

---

[4] We will use e_l for the signed local error and then take the norm $||e_l||$ explicitly when needed. Similarly, e will be the signed global error, with norm $||e||$.

[5] If the method is unstable, the error grows without bound.

2.8 Local Truncation Error vs. Local Error

so if the local error is $(p+1)$th order, the global error is pth order. The actual propagation of errors is a bit more complicated, but the relationship between the orders of the local and global errors remains the same. We will make the argument precise in Sect. 3.10 in the proof of the Equivalence Theorem for ODEs.

The local error $e_l = u(t_{n+1}) - u_{n+1}$, where u_{n+1} approximates the exact solution $u(t_{n+1})$, for a timestep from t_n to t_{n+1} for the initial value problem (2.28) can be calculated quickly by calculating the *local truncation error* (LTE).

The one-step discretized version of the initial value problem (2.28) can be written as

$$u_{n+1} = u_n + \Delta t \, \Phi(u_n, u_{n+1}, \Delta t). \tag{2.30}$$

We define the local truncation error LTE $= \Delta t \, \tau$ for the initial value problem (2.28) by

$$u(t_n + \Delta t) = u(t_n) + \Delta t \, \Phi(u(t_n), u(t_n + \Delta t), \Delta t) + \Delta t \, \tau \tag{2.31}$$

where τ is the global truncation error. In other words, the LTE is the amount by which the exact solution fails to satisfy the discrete method after one timestep.

Now we will prove the LTE equals the actual signed local error e_l after one step of Δt to leading order in Δt, and therefore the global truncation error τ, assuming stability, is of the same order of accuracy as the error e. Thus a one-step method is pth-order accurate if the global truncation error $\tau \sim \Delta t^p$ or equivalently if the LTE $\sim \Delta t^{p+1}$.

Assume we have exact initial data $u(t_n) = u_n$ at t_n. The LTE equals the local error to leading order when Φ in the one-step method is Lipschitz continuous:[6]

$$||\Phi(u(t_{n+1})) - \Phi(u_{n+1})|| \leq C ||u(t_{n+1}) - u_{n+1}|| \tag{2.32}$$

for some positive constant C. The local error going from t_n to t_{n+1} is

$$e_l = u(t_{n+1}) - u_{n+1} = \Delta t \, (\Phi(u(t_{n+1})) - \Phi(u_{n+1})) + \text{LTE}. \tag{2.33}$$

Using the Lipschitz continuity condition,

$$||e_l - \text{LTE}|| \leq C \Delta t \, ||u(t_{n+1}) - u_{n+1}|| = C \Delta t \, ||e_l|| \tag{2.34}$$

so e_l equals the LTE to leading order in Δt.

---

[6] For example, $|x|$ is Lipschitz but not continuously differentiable; $\sqrt{x}$ is continuous but not Lipschitz at $x = 0$. For functions over a closed, bounded subset of the real line: continuously differentiable $\subset$ Lipschitz continuous $\subset$ continuous.

Problems

2.1 Suppose a double precision floating point number is stored on a computer using 64 bits in the following way: sign 1 bit, exponent 8 bits, and mantissa 55 bits. A given real number r is written as

$$r = \pm m 2^n$$

where the mantissa m satisfies $\frac{1}{2} \leq m < 1$ and $-128 \leq n \leq 127$. Give the following numbers both as a power of 2 and in base 10 scientific notation:

(a) What is the machine epsilon ϵ_M (take the leading 1 in m to be a phantom)?
(b) What is the largest positive number *realmax* that can be stored?
(c) What is the smallest positive number *realmin* that can be stored?

Hint: The answers for the IEEE Standard would be (a) $\epsilon_M = 2^{-52} \approx 2.22 \times 10^{-16}$, (b) $realmax = (2 - \epsilon_M) 2^{1023} \approx 2^{1024} \approx 1.8 \times 10^{308}$, (c) $realmin = 2^{-1022} \approx 2.2 \times 10^{-308}$.

2.2 (a) Verify that the three-point central difference formulas for df/dx and $d^2 f/dx^2$ are second-order accurate.
(b) Verify that the one-sided difference formula

$$\frac{du}{dt} \approx \frac{u_{n+1} - u_n}{\Delta t}$$

is first-order accurate. Note that when combined with the trapezoidal rule (TR) ODE or PDE method, the overall method is second order.

2.3 Verify that the approximation

$$\left(\frac{df}{dx}\right)_i \approx \frac{1}{\Delta x}\left[-\frac{1}{12} f_{i+2} + \frac{2}{3} f_{i+1} - \frac{2}{3} f_{i-1} + \frac{1}{12} f_{i-2}\right]$$

is fourth-order accurate.

2.4 Show explicitly that the local error $e_l \approx$ LTE for the backward Euler method

$$u_{n+1} = u_n + a \Delta t \, u_{n+1}$$

for the linear ODE $du/dt = au$, assuming exact initial data at t_n ($u(t_n) = u_n$). *Hint:* Calculate the local error $e_l = u(t_{n+1}) - u_{n+1}$ in terms of the LTE.

Chapter 3
Numerical Methods for ODE IVPs

Ordinary differential equations are divided into two classes: *initial value problems* (IVPs), where initial conditions for the solution are specified at some time t_0, and *boundary value problems* (BVPs), where boundary conditions for the solution are specified at the left and right boundaries. This chapter will discuss numerical methods for ODE initial value problems, and the next chapter numerical methods for ODE boundary value problems.

An ODE initial value problem could equivalently be formulated for a solution depending on x, where initial conditions for the solution are specified at some spatial point x_0.

To solve the ODE initial value problem

$$\frac{du}{dt} = f(u), \quad u(t=0) = u_0 \qquad (3.1)$$

we use a one-step method and approximate

$$\frac{du}{dt} \approx \frac{u_{n+1} - u_n}{\Delta t} = \tilde{f}(u_n, u_{n+1}, \Delta t) \qquad (3.2)$$

where $\tilde{f}(u_n, u_{n+1}, \Delta t)$ is an approximation to the right-hand side of the initial value problem (3.1). Unfolding the time derivative, we obtain the standard form of the numerical method:

$$u_{n+1} = u_n + \Delta t \, \tilde{f}(u_n, u_{n+1}, \Delta t). \qquad (3.3)$$

Different numerical methods are defined by different approximations $\tilde{f}$. If $\tilde{f} = f(u_n)$, we have the forward Euler method, which is *explicit* since the right-hand side of (3.3) only involves the old time level value u_n. If $\tilde{f}$ involves u_{n+1}, the method is called *implicit*. Backward Euler, where $\tilde{f} = f(u_{n+1})$, and TR, where

Table 3.1 Numerical methods for ODE initial value problems. Stability is for $du/dt = au, a < 0$

Method	Type	Order	Stability
forward Euler	Explicit	First	$\Delta t \leq 2/|a|$
backward Euler	Implicit	First	L-stable
TR	Implicit	Second	A-stable
MMR	Implicit	Second	A-stable
predictor-corrector	Explicit	Second	$\Delta t \leq 2/|a|$
RK2	Explicit	Second	$\Delta t \leq 2/|a|$
RK4	Explicit	Fourth	$\Delta t \leq 2.79/|a|$
TRBDF2	Implicit	Second	L-stable

$\tilde{f} = \frac{1}{2}(f(u_n) + f(u_{n+1}))$, are examples of implicit methods. Table 3.1 lists the numerical methods we will discuss in this chapter.

We will develop time-dependent numerical methods for deterministic dynamical systems and analyze the geometry of the phase space of solution trajectories, including chaotic solution behavior and strange attractors in the Shaw oscillator and Lorenz equations. At the end of the chapter, we prove the Equivalence Theorem for ODE initial value and boundary value problems.

See Iserles' *A First Course in the Numerical Analysis of Differential Equations* [22] and Butcher's *Numerical Methods for Ordinary Differential Equations* [5] for general introductions to solving ordinary differential equations.

3.1 Forward and Backward Euler Methods

The *forward Euler* method for $du/dt = f(u)$ is

$$u_{n+1} = u_n + \Delta t \, f(u_n). \tag{3.4}$$

Forward Euler is an explicit method and is first-order accurate and conditionally stable.

For the linear case $du/dt = au$, the exact solution is $u(t) = u_0 e^{at}$. The growth factor for forward Euler is defined by

$$u_{n+1} = (1 + a\Delta t)u_n = Gu_n. \tag{3.5}$$

The exact growth factor $H = e^{a\Delta t}$. Note that $G = e^{a\Delta t}$ to first order in Δt (*consistency*). For n timesteps,

$$u_n = G^n u_0 = (1 + a\Delta t)^n u_0 \to e^{na\Delta t} u_0 = e^{at_n} u_0 = u(t_n) \tag{3.6}$$

as $n \to \infty$ and $\Delta t \to 0$ with $n\Delta t = t_n$ held fixed. We have proved u_n *converges* to the exact solution $u(t_n)$ in this limit. (Henceforth we will prove that a numerical

3.1 Forward and Backward Euler Methods

method is consistent and stable; then convergence follows from the Equivalence Theorem.)

To analyze *stability*, we consider the model problem

$$\frac{du}{dt} = au, \quad a < 0 \quad \text{(model problem)} \tag{3.7}$$

because then the exact solution $u(t) = e^{-|a|\Delta t}u_0$ decays. The computed solution $u_n = G^n u_0$ must not grow for $a < 0$, so stability requires $|G| \leq 1$. For forward Euler, we have $-1 \leq 1 + a\Delta t = 1 - |a|\Delta t \leq 1$ or $\Delta t \leq 2/|a|$ for stability. The neutrally stable case in which $|G| = 1$ requires caution, since the numerical solution may be totally inaccurate (see Problem 3.8).

To find the order of accuracy for the full nonlinear problem (3.1), we calculate the global truncation error τ:

$$u(t + \Delta t) = u + \Delta t\, f(u) + \Delta t\, \tau \tag{3.8}$$

where t is a shorthand for t_n and u is a shorthand for $u(t_n)$. Taylor expanding (and using $f(u) = du/dt \equiv u'$), we get

$$u + \Delta t\, u' + \frac{1}{2}\Delta t^2 u'' + \cdots = u + \Delta t\, u' + \Delta t\, \tau \tag{3.9}$$

$$\tau = \frac{1}{2}\Delta t\, u'' + O(\Delta t^2) \tag{3.10}$$

which shows forward Euler is first-order accurate.

The *backward Euler* method for $du/dt = f(u)$ is

$$u_{n+1} = u_n + \Delta t\, f(u_{n+1}). \tag{3.11}$$

Backward Euler is an implicit method, first-order accurate, and A-stable and L-stable (see Problems 3.1 and 3.2).

For linear systems of ODEs $u' = Au$, the model problem (3.7) is replaced by $u' = \lambda u$, where λ is an eigenvalue of A. For nonlinear systems of ODEs $u' = f(u)$, we linearize and examine the eigenvalues λ of the Jacobian $J = \partial f/\partial u$ in (2.23) $\delta u' = J\delta u$ when $\delta u' = \lambda \delta u$. Note that λ will in general be complex even for real matrices A or J. Thus, we can plot the stability region where the complex growth factor $G(z)$, $z = \lambda \Delta t$, has modulus $|G(z)| \leq 1$ in the complex plane. The stability regions for forward and backward Euler, TR, and TRBDF2 are plotted in Fig. 3.1. A-stability corresponds now to a stability region that includes the entire left half-plane $\mathcal{R}\{z\} \leq 0$. Thus backward Euler is A-stable, but forward Euler is only conditionally stable.

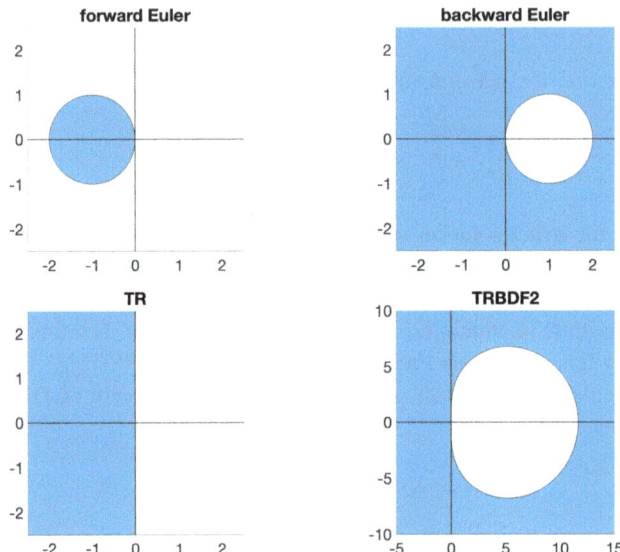

Fig. 3.1 Stability regions (*shaded*) in the complex plane showing only conditional stability for the forward Euler method, and A-stability for the backward Euler, trapezoidal rule, and TRBDF2 ($\gamma = 2 - \sqrt{2}$) methods. Backward Euler and TRBDF2 are also L-stable

3.2 Trapezoidal Rule Method

The *trapezoidal rule (TR)* method for $du/dt = f(u)$ is

$$u_{n+1} = u_n + \frac{\Delta t}{2}(f(u_n) + f(u_{n+1})). \tag{3.12}$$

TR is implicit, second-order accurate, and A-stable but not L-stable.

Stability: The growth factor for TR for $du/dt = au$ is

$$u_{n+1} = \frac{1 + \frac{1}{2}a\Delta t}{1 - \frac{1}{2}a\Delta t} u_n = G u_n. \tag{3.13}$$

For the model problem ($a < 0$),

$$-1 \leq \frac{1 - \frac{1}{2}|a|\Delta t}{1 + \frac{1}{2}|a|\Delta t} = \frac{1 - \Delta}{1 + \Delta} \leq 1, \quad \Delta = \frac{1}{2}|a|\Delta t > 0 \tag{3.14}$$

is always true, independent of Δt. TR is not L-stable since

$$\lim_{\Delta t \to \infty} G = -1. \tag{3.15}$$

Consistency: The LTE ($= \Delta t \, \tau$) is given by

$$u(t + \Delta t) = u + \frac{\Delta t}{2}(f(u) + f(u(t + \Delta t))) + \Delta t \, \tau. \tag{3.16}$$

Taylor expanding (and using $f(u) = u'$), we get

$$u + \Delta t \, u' + \frac{1}{2}\Delta t^2 u'' + \frac{1}{6}\Delta t^3 u''' + \cdots = u + \Delta t \, u' + \frac{1}{2}\Delta t^2 u'' + \frac{1}{4}\Delta t^3 u''' + \cdots + \Delta t \, \tau \tag{3.17}$$

$$\tau = -\frac{1}{12}\Delta t^2 u''' + O(\Delta t^3) \tag{3.18}$$

which shows TR is second-order accurate.

The stability region for TR is shown in Fig. 3.1, demonstrating graphically that TR is A-stable.

3.3 Predictor-Corrector Methods

Predictor-corrector methods are frequently used in fluid dynamical computations.

The simplest predictor-corrector method is the second-order (explicit) Adams-Bashforth-Moulton method. For $du/dt = f(t, u)$, the predictor step is a forward Euler extrapolation:

$$\tilde{u}_{n+1} = u_n + \Delta t \, f(t_n, u_n) \tag{3.19}$$

which is followed by a second-order accurate corrector step:

$$u_{n+1} = u_n + \frac{\Delta t}{2}\left(f(t_n, u_n) + f(t_{n+1}, \tilde{u}_{n+1})\right). \tag{3.20}$$

Note the similarities and differences with the TR (3.12), RK2 (3.23), and MMR (Problem 3.3) methods. Problem 3.4 demonstrates that this predictor-corrector method is second-order accurate and conditionally stable.

A fourth-order (explicit) Adams-Bashforth-Moulton method can be obtained by fitting the predictor and corrector steps by cubic polynomials in f. The predictor step fits a cubic polynomial through the four most recent points in the derivative picture, extrapolates it, and integrates under the cubic (see Problem 3.5):

$$\tilde{u}_{n+1} = u_n + \frac{\Delta t}{24}\left(55 f(t_n, u_n) - 59 f(t_{n-1}, u_{n-1})\right.$$
$$\left. + 37 f(t_{n-2}, u_{n-2}) - 9 f(t_{n-3}, u_{n-3})\right). \tag{3.21}$$

The corrector step also fits a cubic polynomial through the four most recent points in the derivative picture and integrates under the cubic:

$$u_{n+1} = u_n + \frac{\Delta t}{24}\left(9f(t_{n+1}, \tilde{u}_{n+1})\right.$$
$$\left. + 19f(t_n, u_n) - 5f(t_{n-1}, u_{n-1}) + f(t_{n-2}, u_{n-2})\right). \quad (3.22)$$

However, this step does not involve extrapolation, since we now know $f(t_{n+1}, \tilde{u}_{n+1})$.

For both predictor-corrector methods, the timestep is adjusted dynamically based on the methods of Sect. 3.6, approximating $||e_l|| \approx ||u_{n+1} - \tilde{u}_{n+1}||$.

3.4 Runge-Kutta Methods

The *RK2* method for $du/dt = f(u)$ is

$$u_{n+1} = u_n + \Delta t\, f\left(u_n + \frac{1}{2}\Delta t\, f(u_n)\right). \quad (3.23)$$

RK2 is explicit, second-order accurate, and conditionally stable: $\Delta t \leq 2/|a|$ for $f(u) = au$, $a < 0$ (see Problem 3.6).

If $f = f(t, u)$ depends on u and explicitly also on t (nonautonomous form), then the RK2 method is

$$u_{n+1} = u_n + \Delta t\, f\left(t_n + \frac{1}{2}\Delta t,\, u_n + \frac{1}{2}\Delta t\, f(t_n, u_n)\right). \quad (3.24)$$

Note that a nonautonomous ODE $du/dt = f(t, u)$ can always be transformed into a pair of autonomous ODEs by setting $v = t$:

$$\begin{bmatrix} du/dt \\ dv/dt \end{bmatrix} = \begin{bmatrix} f(v, u) \\ 1 \end{bmatrix}, \quad \begin{bmatrix} u(t_0) \\ v(t_0) \end{bmatrix} = \begin{bmatrix} u_0 \\ t_0 \end{bmatrix}. \quad (3.25)$$

The *RK4* method for $du/dt = f(u)$ is

$$u_{n+1} = u_n + \frac{\Delta t}{3}(K_1 + 2K_2 + 2K_3 + K_4) \quad (3.26)$$

with

$$K_1 = \frac{1}{2}f(u_n), \quad K_2 = \frac{1}{2}f(u_n + \Delta t\, K_1),$$

3.5 Harmonic Oscillator

$$K_3 = \frac{1}{2}f(u_n + \Delta t\, K_2), \quad K_4 = \frac{1}{2}f(u_n + 2\Delta t\, K_3). \quad (3.27)$$

RK4 is explicit, fourth-order accurate, and conditionally stable (see Problem 3.7).

3.5 Harmonic Oscillator

As a practical application, we will solve the dynamical equation for the motion of a damped harmonic oscillator:

$$\frac{d^2 y}{dt^2} + 2p\frac{dy}{dt} + \omega_0^2 y = 0 \quad (3.28)$$

where for frictional damping $p > 0$ and where ω_0 is the natural angular frequency of the oscillator without friction, with initial conditions $y(0) = y_0$, $y'(0) = v_0$.

To convert this second-order ODE to a first-order system of ODEs, set $u(t) = y(t)$ and $v(t) = y'(t)$. Then

$$\frac{du}{dt} = v \quad (3.29)$$

$$\frac{dv}{dt} = -2pv - \omega_0^2 u \quad (3.30)$$

with initial conditions $u(0) = y_0$, $v(0) = v_0$. Defining $w(t)$ as the column vector $(u(t), v(t))$, we can write the two first-order ODEs in vector form as

$$\frac{dw}{dt} = \frac{d}{dt}\begin{bmatrix} u \\ v \end{bmatrix} = f(w) = \begin{bmatrix} v \\ -2pv - \omega_0^2 u \end{bmatrix} \quad (3.31)$$

with initial conditions $w(0) = w_0 = (u_0, v_0)$. Since the damped harmonic oscillator ODE is linear,

$$f(w) = \begin{bmatrix} 0 & 1 \\ -\omega_0^2 & -2p \end{bmatrix} w = Aw. \quad (3.32)$$

Modern methods for time-dependent problems always employ a variable Δt adjusted dynamically based on an estimate of the local error. Our model program **rk2dyn.m** implements this idea. The computed and exact solutions for the damped harmonic oscillator are displayed in Fig. 3.2.

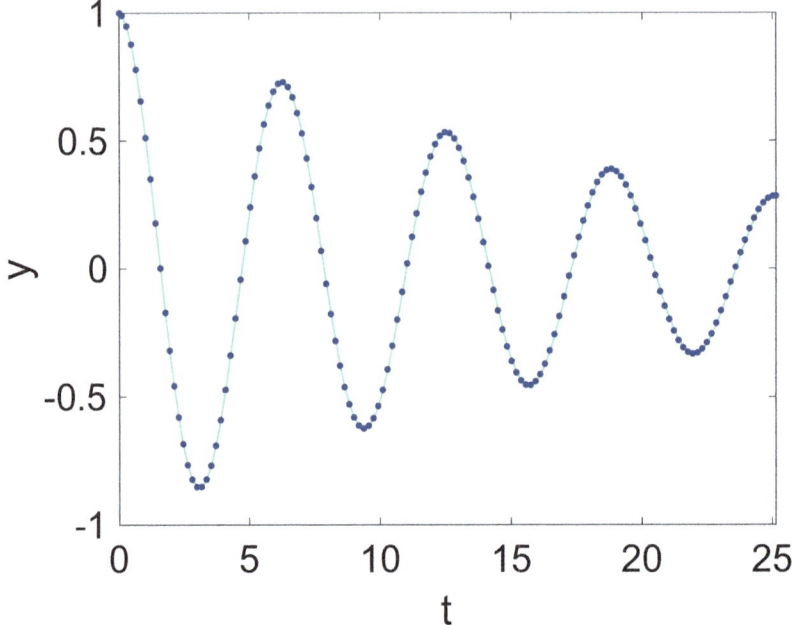

Fig. 3.2 Harmonic oscillator solution with frictional damping computed using **rk2dyn.m** (*dots*) with adaptive timestep, along with the exact solution (*curve*)

3.6 Dynamic Timestep

For time-dependent problems (ODE initial value problems and parabolic, hyperbolic, and mixed type PDEs), the timestep Δt should always be adjusted dynamically.

For hyperbolic PDEs, explicit methods are recommended, and the timestep is determined dynamically by the CFL condition for stability

$$\Delta t = r \frac{\Delta x}{\max\{|\lambda|\}} \tag{3.33}$$

where the Courant number r satisfies[1] $0 < r \leq 1$ and $\max\{|\lambda|\}$ is the maximum characteristic speed in magnitude on the grid at the beginning of the timestep. Typically r is chosen to be $0.1 \leq r \leq 0.9$.

There are three popular methods for adjusting Δt for ODE initial value problems and parabolic PDEs based on an estimate of the local error: Two computed solutions at t_{n+1} can be compared to estimate e_l, using either the classic Milne strategy below

---

[1] For some hyperbolic schemes $0 < r \leq \frac{1}{2}$, etc., for stability.

3.6 Dynamic Timestep

Fig. 3.3 Time levels for dynamic timestep

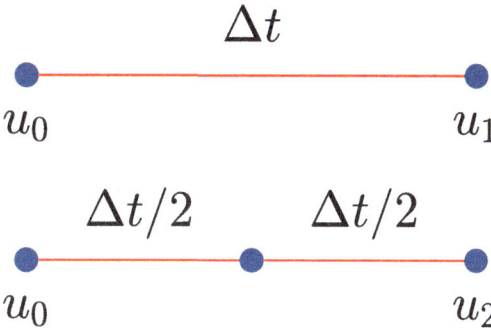

or using two numerical solutions computed with different methods (e.g., RK4 and RK5); or a divided-difference formula can be used in the TRBDF2 method.

Suppose we have two estimates (Fig. 3.3) of the exact solution $\bar{u}$ at time level $n+1$, based on initial data u_0 at time level n: u_1 calculated with a single step of Δt, and u_2 calculated with two steps of $\Delta t/2$. Then using a technique introduced by Milne, we can estimate the local error $e_l = \bar{u} - u_1 \sim u_2 - u_1$. In fact, for a pth-order method,

$$e_l = \bar{u} - u_1 \approx C \Delta t^{p+1} \tag{3.34}$$

$$u_2 \approx \bar{u} - 2 e_l(\Delta t/2) \tag{3.35}$$

$$u_2 - u_1 \approx e_l(\Delta t) - 2 e_l(\Delta t/2) \approx \left(1 - \frac{1}{2^p}\right) e_l \tag{3.36}$$

$$e_l \approx \frac{2^p}{2^p - 1}(u_2 - u_1). \tag{3.37}$$

To dynamically adjust Δt, we monitor the estimate of the local error e_l. Define the local ratio r_l, which should be less than or on the order of 1:

$$r_l = \frac{||e_l||}{\epsilon_R ||u_n|| + \epsilon_A} \lesssim 1 \tag{3.38}$$

where u_n is the initial data at time level n, ϵ_R specifies the relative error tolerance, and ϵ_A specifies an absolute error tolerance (*effective zero*), so that the denominator of r_l never equals zero. In other words, we are requiring

$$||e_l|| \lesssim \epsilon_R ||u_n|| + \epsilon_A. \tag{3.39}$$

Our dynamic timestep algorithm is

- If $r_l \leq 2$, accept the provisional solution $u_{n+1} = u_2$ and set the new $\Delta t \leftarrow \Delta t / r_l^{1/(p+1)}$.
- If $r_l > 2$, redo the timestep with $\Delta t \leftarrow \Delta t / r_l^{1/(p+1)}$ or $\Delta t \leftarrow \Delta t/2$.

The scaling of Δt maintains $||e_l|| \approx \epsilon_R ||u||$, since we are extrapolating that at time level $n+2$

$$r_{l,new} \approx \frac{||e_{l,new}||}{\epsilon_R ||\tilde{u}|| + \epsilon_A} \approx \frac{||e_{l,new}||}{||e_l||} r_l \approx \frac{\Delta t_{new}^{p+1}}{\Delta t^{p+1}} r_l = 1 \qquad (3.40)$$

for

$$\Delta t_{new} = \frac{\Delta t}{r_l^{1/(p+1)}} \qquad (3.41)$$

where $\tilde{u} \approx u_n \approx u_{n+1}$.

Figure 3.4 illustrates the dynamic timestepping algorithm with RK2/RK2 for the nonlinear van der Pol oscillator described in Sect. 3.7.2. Notice how the timestep opens up when the solution evolves slowly, while it closes down when the solution

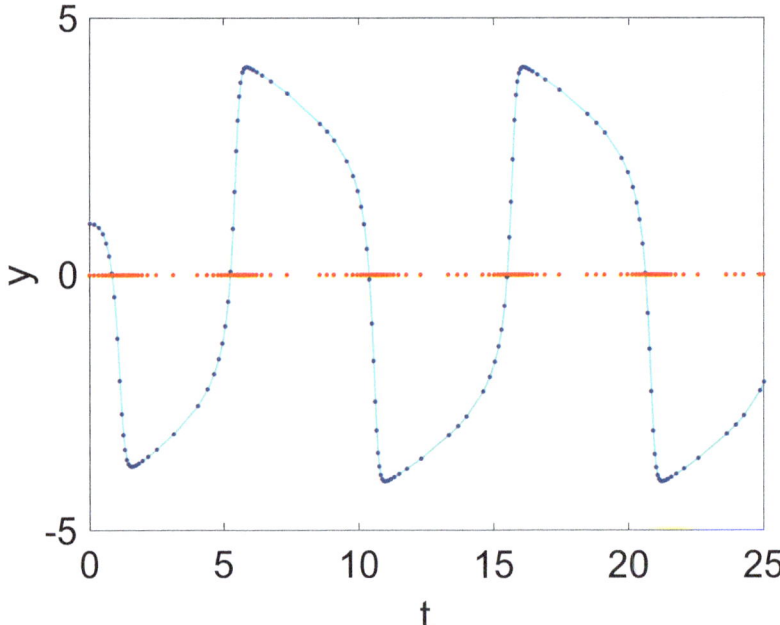

Fig. 3.4 Van der Pol oscillator solution computed using **rk2dyn.m** with adaptive timestep: solution values (*dots* connected by the *curve*), and time levels (*dots* along the time axis). Notice how the timestep opens up in the smooth regions of the solution

3.6 Dynamic Timestep

Table 3.2 Global relative errors $e_R = |u(t_n) - u_n|/|u(t_n)|$ for van der Pol oscillator simulations using MATLAB's **ode45** RK4/RK5 solver for different values of ϵ_R and $\epsilon_A = \epsilon_M$. Here $u(t_n = 10) = 1.3855149156118856$ to 17 digits from the *Mathematica* ODE solver

ϵ_R	Steps	e_R
10^{-3}	265	5.4×10^{-4}
10^{-6}	745	3.7×10^{-6}
10^{-9}	2757	2.6×10^{-9}
10^{-12}	10,845	2.2×10^{-12}

varies rapidly. Table 3.2 demonstrates how the local relative error criterion ϵ_R influences the global relative error e_R for MATLAB's **ode45** solver, where the two different numerical solutions at the new time level are produced with RK4 (u_1) and RK5 (u_2).

The program **rk2dyn.m** solves the damped harmonic oscillator, van der Pol, and Shaw oscillator problems, as well as the Lorenz equations, with a variable Δt (see Sect. 3.7 for a description of the nonlinear dynamical systems). To see the timestep information in **rk2dyn.m**, turn on verbose mode by setting VERBOSE = 1 at the top of the program.

For **rk2dyn.m**, try

```
rk2dyn([1; -0.05],8*pi,1)    % damped harmonic oscillator
rk2dyn([1; 0],50,2)          % van der Pol oscillator
rk2dyn(1,2,3)                % singularity detection
rk2dyn([0; 1; 0],40,4)       % Lorenz equations
```

where singularity detection is for $y' = y^2$, $y(0) = 1$, $y(t) = 1/(1-t)$.

For the initial Δt_0, recall we want $||e_l|| \approx \epsilon_R ||u||$. A natural timescale for the initial value problem $du/dt = f(u)$ is

$$\bar{t} = \frac{||u_0|| + \frac{\epsilon_A}{\epsilon_R}}{||f_0|| + \frac{\epsilon_A}{\epsilon_R}}. \tag{3.42}$$

Set $\Delta t_0 = \epsilon_R^{1/(1+p)} \bar{t}$ for a pth-order method. Then

$$||e_l|| = C \Delta t_0^{p+1} ||u_0^{(p+1)}|| = C \epsilon_R \bar{t}^{p+1} ||u_0^{(p+1)}|| \sim \epsilon_R ||u_0||. \tag{3.43}$$

For implicit methods for parabolic PDEs like linear ($D(u) = const$) and nonlinear diffusion $u_t = (D(u)u_x)_x$, the initial timestep can be taken to be

$$\Delta t_0 = \frac{\sqrt{N} \Delta x^2}{2 D_{max}} \tag{3.44}$$

where N is the number of Δx and D_{max} is the maximum value of $D(u)$ on the grid at t_0.

3.7 Deterministic Dynamical Systems

We will always consider *deterministic* dynamical systems $dw/dt = f(w)$ with a continuously differentiable (and therefore Lipschitz continuous) $f(w)$: i.e., the Jacobian $J = \partial f/\partial w$ exists and is continuous. The relevant theorem on existence and uniqueness of solutions is due to Picard and Lindelöf:

Theorem *Picard-Lindelöf Theorem: A unique solution to the initial value problem $dw/dt = f(t, w)$, $w(t_0) = w_0$, exists for $t_0 - \epsilon \le t \le t_0 + \epsilon$ for some $\epsilon > 0$ if f is uniformly Lipschitz continuous in w and continuous in t.*

The ODE initial value problem

$$\frac{dy}{dt} = y^{\frac{1}{3}}, \quad y(0) = 0 \tag{3.45}$$

is an example of a nondeterministic problem that has three possible solutions for $t > 0$:

$$y(t) = \begin{cases} 0 \\ \pm \left(\frac{2}{3}t\right)^{\frac{3}{2}} \end{cases} \tag{3.46}$$

This occurs because $y^{\frac{1}{3}}$ is continuous (guarantees existence by the Peano Existence Theorem) but not Lipschitz continuous at $y = 0$ (fails the Picard-Lindelöf uniqueness test).

Poincaré introduced the idea of analyzing the geometry of the phase space of solution trajectories to a deterministic dynamical system. We will illustrate this type of analysis in the context of attractors (limit points, limit cycles, limit tori, and strange attractors) in phase space.

3.7.1 Nonlinear Pendulum

The nonlinear pendulum obeys the initial value problem

$$u'' + \gamma u' + \sin(u) = 0, \quad u(0) = u_0, \quad u'(0) = v_0 \tag{3.47}$$

where $u(t)$ is the angular displacement of the pendulum from the vertical and γ is the coefficient of frictional damping. The modern approach is to rewrite the second-order ODE (3.47) as a system of two first-order equations

$$y' = \begin{bmatrix} y_1' \\ y_2' \end{bmatrix} = f(y) = \begin{bmatrix} y_2 \\ -\sin(y_1) - \gamma y_2 \end{bmatrix}. \tag{3.48}$$

3.7 Deterministic Dynamical Systems

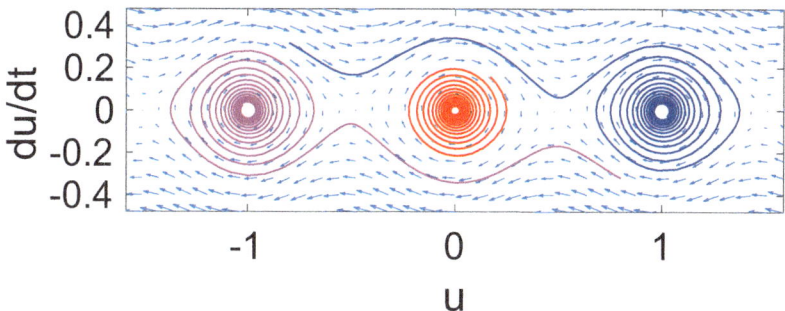

Fig. 3.5 Phase space diagram and direction field for the nonlinear pendulum with frictional damping computed using **ode45**. Note the three limit points in the diagram. u and du/dt have been divided by 2π

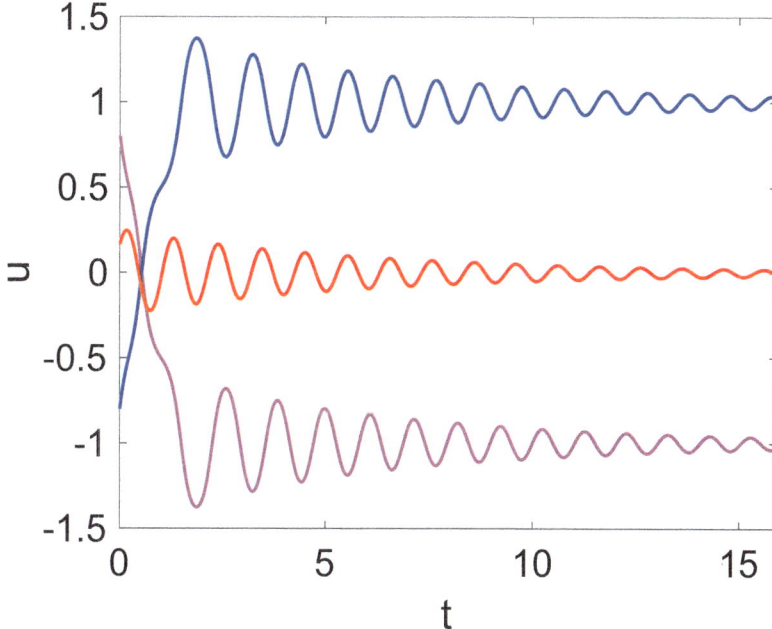

Fig. 3.6 Solutions $u(t)$ for the nonlinear pendulum with frictional damping computed using **ode45**. u and t have been divided by 2π

Three solutions computed with **ode45** are shown in Figs. 3.5 and 3.6.

3.7.2 Van der Pol Oscillator

The van der Pol equations, originally derived for a triode oscillator, are

$$\frac{d^2y}{dt^2} + (y^2 - \epsilon)\frac{dy}{dt} + y = 0, \quad y(0) = 1, \quad \frac{dy}{dt}(0) = 0. \tag{3.49}$$

For $y^2 > \epsilon$, the $(y^2 - \epsilon)y'$ term acts as (nonlinear) friction causing $y \to 0$, while for $y^2 < \epsilon$, the $(y^2 - \epsilon)y'$ term acts as negative friction causing $y \to \pm\infty$. The competition between the two (positive and negative) frictional terms produces a closed loop limit cycle in phase space.

Reexpress the second-order ODE as two first-order ODEs with $u = y$, $v = dy/dt$:

$$\frac{du}{dt} = v \tag{3.50}$$

$$\frac{dv}{dt} = (\epsilon - u^2)v - u \tag{3.51}$$

with initial conditions $u(0) = 1$, $v(0) = 0$. The attractor in phase space exhibiting an asymptotically periodic orbit is shown in Fig. 3.7 for $\epsilon = 4$ (take $t_f = 40$).

3.7.3 Shaw Oscillator

In the van der Pol equations, set $u \to v$ and $v \to -u$, and then add the sinusoidal forcing term to the (wrong!) new dv/dt equation to obtain (up to constants) the Shaw oscillator equations:

$$\frac{du}{dt} = 0.7v + 10u(0.1 - v^2) \tag{3.52}$$

$$\frac{dv}{dt} = -u + 0.25\sin(1.57w) \tag{3.53}$$

$$\frac{dw}{dt} = 1. \tag{3.54}$$

With these parameters, the solution has a strange attractor with fractal dimension ≈ 2.6. For initial conditions, take $u(0) = -0.73$, $v(0) = 0$, $w(0) = 0$ (w is time t), with $t_f = 100$ (see Fig. 3.8). We can make w periodic with period $2\pi/1.57$ by identifying $w + 2m\pi/1.57 \equiv w$, $m = \pm 1, \pm 2, \ldots$ With w periodic, the 3D attractor

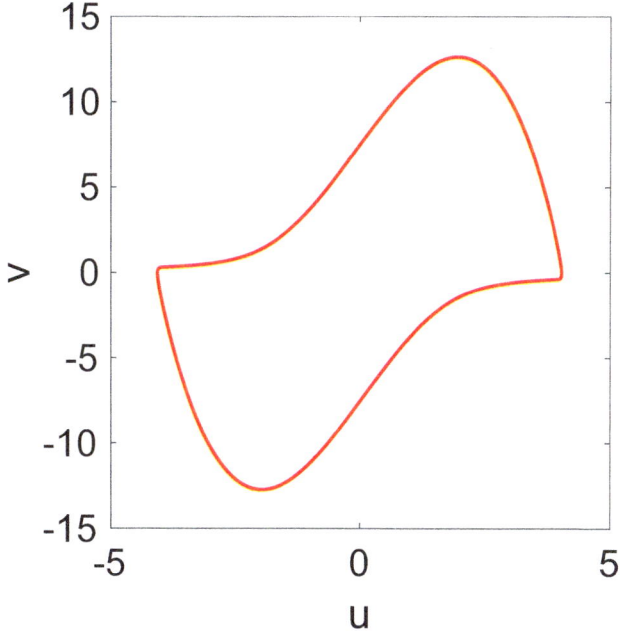

Fig. 3.7 Van der Pol oscillator limit-cycle attractor in the u-v phase space computed with **rk2dyn.m**. The transient solution points have been discarded to obtain the large time attractor

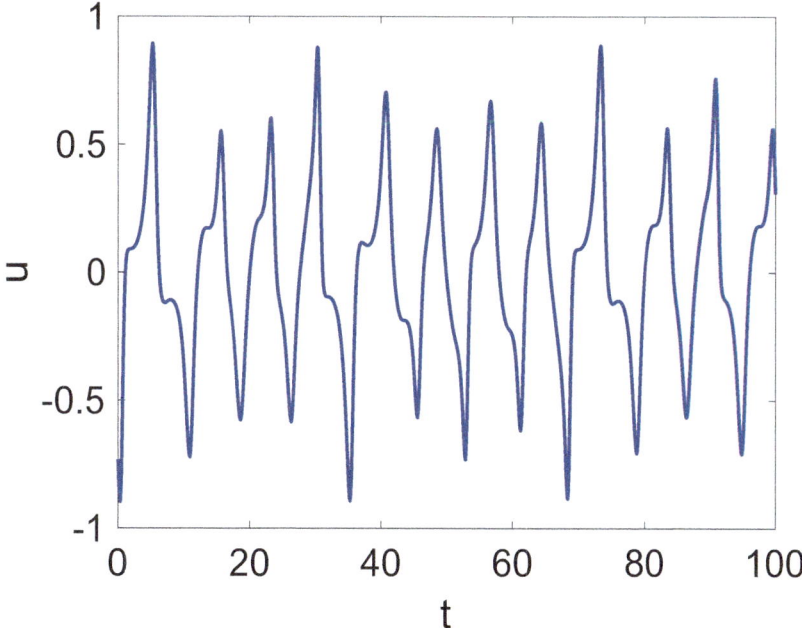

Fig. 3.8 Shaw oscillator solution computed with **ode45**

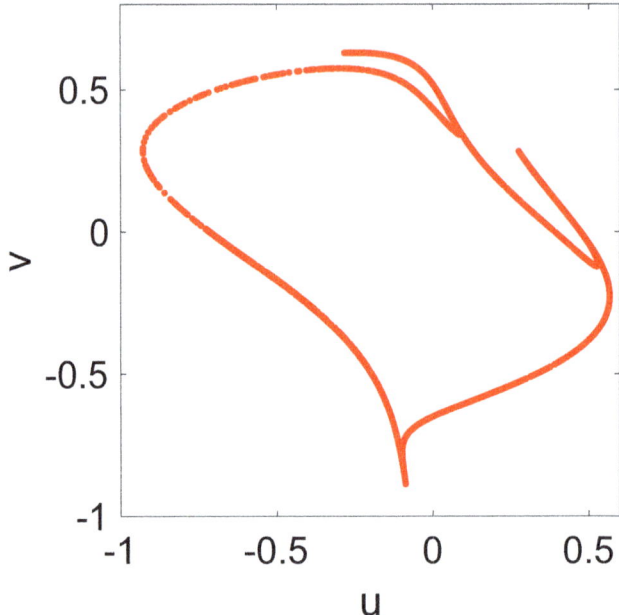

Fig. 3.9 Poincaré slice of Shaw oscillator strange attractor in the u-v phase space computed with **rk2.m**. The transient solution points have been discarded. The cusps are indicative of a strange (fractal dimension) attractor

in the u-v-w phase space is a torus with cusps. The Poincaré slice (Fig. 3.9) of the attractor in the u-v phase space indicates a strange (fractal dimension) attractor.

3.7.4 Lorenz Equations

The Lorenz equations (1963) [28] model vertical counter-rotating vortices in Rayleigh-Bénard convection:

$$\frac{dx}{dt} = \sigma(y - x) \tag{3.55}$$

$$\frac{dy}{dt} = x(r - z) - y \tag{3.56}$$

$$\frac{dz}{dt} = xy - bz. \tag{3.57}$$

3.7 Deterministic Dynamical Systems

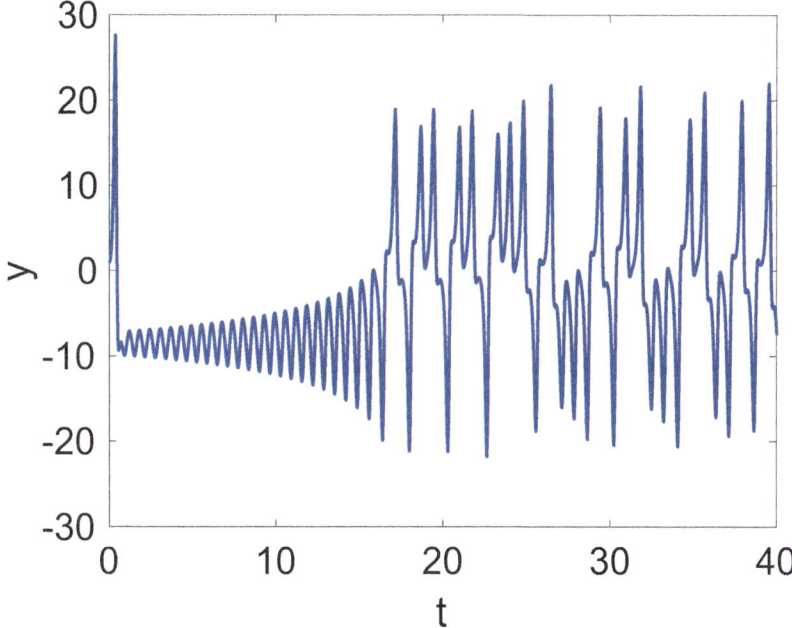

Fig. 3.10 Lorenz equations solution $y(t)$ computed with **ode45**

For a Rayleigh-Bénard vortex, $x(t) \sim$ the angular velocity, $y(t \to \infty) \sim T$ at the middle right edge, and $z(t \to \infty) \sim T$ at the bottom. Here σ is the Prandtl number[2] (10 is appropriate for cold water; too high for air), r is the Rayleigh number,[3] and b is the aspect ratio of the vortex cell.

For $\sigma = 10$, $r = 28$, and $b = 8/3$, the solution (Fig. 3.10) has a strange attractor (Fig. 3.11) with fractal dimension ≈ 2.06. For initial conditions, take $x(0) = 0$, $y(0) = 1$, and $z(0) = 0$, with $t_f = 30$.

3.7.5 Stability of Equilibria

The equilibria w_{eq} of an initial value problem system $dw/dt = f(w)$ are given by $f(w_{eq}) = 0$. To analyze the stability of the equilibria, make a small perturbation: $w(t) = w_{eq} + \delta w(t)$. Linearizing with respect to δw, we obtain

---

[2] Dimensionless ratio ν/κ of momentum diffusivity (kinematic viscosity) to thermal diffusivity.
[3] Dimensionless number describing heat flow—if the Rayleigh number is below (above) a critical value, heat transfer is dominated by conduction (convection).

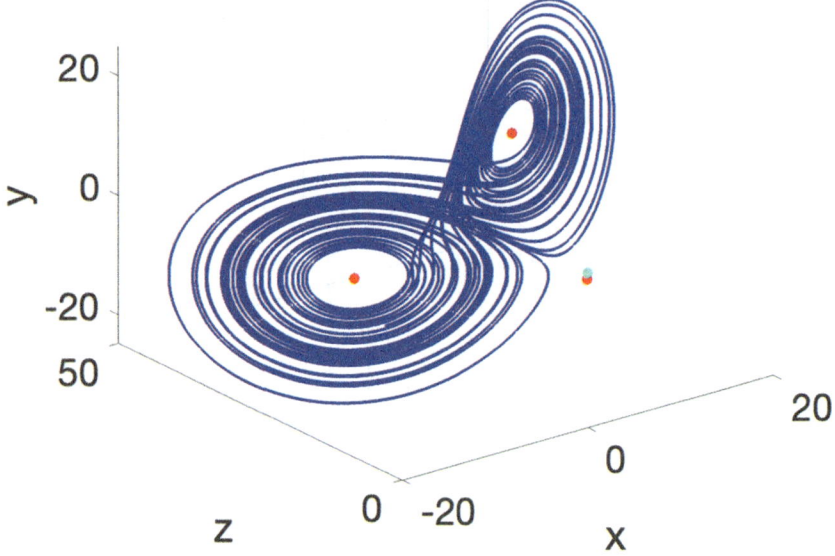

Fig. 3.11 Lorenz equations strange attractor in the x-y-z phase space computed with **ode45**. The transient solution points have been discarded. The *dark dots* are the three equilibria, and the *light dot* represents the initial conditions

$$\frac{d\delta w}{dt} = f(w_{eq} + \delta w) \approx J\,\delta w, \quad J = \left.\frac{\partial f}{\partial w}\right|_{w_{eq}}. \qquad (3.58)$$

Then set $\delta w(t) = e^{\lambda t}\tilde{w}$ where $\tilde{w}$ is a constant vector. We obtain an eigenvalue problem

$$\lambda \delta w = J\,\delta w. \qquad (3.59)$$

If $\mathfrak{R}\{\lambda_j\} < 0$ for all j, then the equilibrium is stable. If $\mathfrak{R}\{\lambda_j\} > 0$ for any j, then the equilibrium is unstable.

Write the Lorenz equations (3.55)–(3.57) as

$$\frac{dw}{dt} = \frac{d}{dt}\begin{bmatrix} x \\ y \\ z \end{bmatrix} = f(w) = \begin{bmatrix} \sigma(y-x) \\ x(r-z) - y \\ xy - bz \end{bmatrix}. \qquad (3.60)$$

There are three equilibria $f(w_{eq}) = 0$:

$$(x, y, z)_{eq} = (\eta, \eta, r-1), \quad (-\eta, -\eta, r-1), \quad (0, 0, 0) \qquad (3.61)$$

where $\eta = \sqrt{b(r-1)}$. The eigenvalues λ_j ($j = 1, 2, 3$) of the Jacobian

$$J = \frac{\partial f}{\partial w} = \begin{bmatrix} -\sigma & \sigma & 0 \\ r-z & -1 & -x \\ y & x & -b \end{bmatrix} \qquad (3.62)$$

evaluated at $J(w_{eq})$ determine the stability of the equilibria. All three equilibria are unstable for the canonical values $\sigma = 10$, $r = 28$, and $b = 8/3$, with the centers of the "butterfly wings" $(\eta, \eta, r-1)$ and $(-\eta, -\eta, r-1)$ slightly unstable and the origin $(0, 0, 0)$ extremely unstable (run **lorenzeigs.m** for an analysis).

3.8 Backward Euler and Newton's Method

When we apply an implicit method to a nonlinear initial value problem $du/dt = f(u)$, the nonlinear discretized equation for u_{n+1} is best solved using Newton's method.

We start with the simple example of applying backward Euler and Newton's method to solving a nonlinear initial value problem; then in the next section, we extend the methodology to TRBDF2.

To integrate $du/dt = f(u)$ from $t = t_n$ to $t_{n+1} = t_n + \Delta t$, the backward Euler formula is

$$u_{n+1} = u_n + \Delta t \, f(u_{n+1}). \qquad (3.63)$$

We linearize $f(u_{n+1})$ in (3.63) iteratively by setting the new iterative solution to

$$u_{n+1}^{(k+1)} = u_{n+1}^{(k)} + \Delta u^{(k)}, \quad u_{n+1}^{(0)} = u_n, \quad k = 0, 1, 2, \ldots \qquad (3.64)$$

and approximating

$$f\left(u_{n+1}^{(k+1)}\right) = f\left(u_{n+1}^{(k)}\right) + \left(\frac{\partial f}{\partial u}\right)_{n+1}^{(k)} \Delta u^{(k)} \qquad (3.65)$$

where $k = 0, 1, 2, \ldots$ labels the Newton iterations. Then, we iterate until the Newton method converges.

The Newton equation for backward Euler is

$$\left[I - \Delta t \left(\frac{\partial f}{\partial u}\right)_{n+1}^{(k)}\right] \Delta u^{(k)} = -u_{n+1}^{(k)} + u_n + \Delta t f\left(u_{n+1}^{(k)}\right) \qquad (3.66)$$

or

$$\left(\frac{\partial R_{\text{BE}}}{\partial u_{n+1}}\right)^{(k)}_{n+1} \Delta u^{(k)} = -R_{\text{BE}} \qquad (3.67)$$

where $R_{\text{BE}} = u_{n+1}^{(k)} - u_n - \Delta t \, f\left(u_{n+1}^{(k)}\right)$ is the residual for (3.63). Note that (3.67) has the symbolic Newton form $R' \Delta u = -R$.

The Jacobian can be evaluated by finite differences in the following way if an analytical Jacobian is not available:

$$J_{ij} = \frac{\partial f_i}{\partial u_j} \approx \frac{\Delta f_i}{\Delta u_j} = \frac{f_i(u + \Delta u) - f_i(u)}{\Delta u_j} \approx \frac{\sum_k \frac{\partial f_i}{\partial u_k} \Delta u_k}{\Delta u_j} = \frac{\frac{\partial f_i}{\partial u_j} \Delta u_j}{\Delta u_j} = \frac{\partial f_i}{\partial u_j} \qquad (3.68)$$

by setting

$$\Delta u_k = \begin{cases} \Delta u_j & \text{if } k = j \\ 0 & \text{if } k \neq j \end{cases} \qquad (3.69)$$

where $i, j, k = 1, 2, \ldots, m$ for m-dimensional u and f and where $\Delta u_j = 10^{-9}$, for example.

3.9 TRBDF2 Method

For *stiff* ODE or PDE initial value problems, TRBDF2 [1] is the method of choice, since it is L-stable and second-order accurate.

Stiffness is a subtle concept. A general, broad formulation is that an initial value problem is stiff if it has two or more widely disparate solution timescales. A more precise, narrower definition is given by Moler (Sect. 7.9 of [29]):

> A problem is stiff if the solution being sought varies slowly, but there are nearby solutions that vary rapidly, so the numerical method must take small steps to obtain satisfactory results.

The best way to understand stiffness is through two examples: (i) the L-stable vs. only A-stable computed solutions of the nonlinear diffusion PDE (Fig. 2.3) and (ii) the flame propagation ODE in Problem 3.16.

To integrate $du/dt = f(u)$ from $t = t_n$ to $t_{n+1} = t_n + \Delta t$ using the TRBDF2 method (Fig. 3.12), first apply the trapezoidal rule (TR) to advance the solution from t_n to $t_{n+\gamma} = t_n + \gamma \Delta t$:

$$u_{n+\gamma} - \gamma \frac{\Delta t}{2} f_{n+\gamma} = u_n + \gamma \frac{\Delta t}{2} f_n \qquad (3.70)$$

3.9 TRBDF2 Method

Fig. 3.12 Time levels for TRBDF2

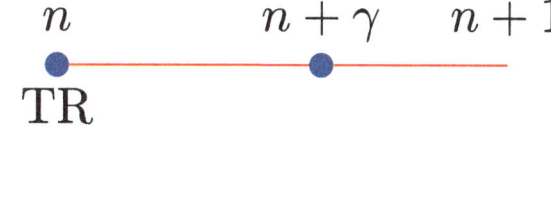

and then use the second-order backward differentiation formula (BDF2)[4] to advance the solution to t_{n+1}:

$$u_{n+1} - \frac{1-\gamma}{2-\gamma}\Delta t\, f_{n+1} = \frac{1}{\gamma(2-\gamma)}u_{n+\gamma} - \frac{(1-\gamma)^2}{\gamma(2-\gamma)}u_n. \tag{3.71}$$

This composite one-step method is second-order accurate and L-stable.

We linearize f_{n+1} in (3.71) (and similarly $f_{n+\gamma}$ in (3.70)) iteratively by setting the new iterative solution to

$$u_{n+1}^{(k+1)} = u_{n+1}^{(k)} + \Delta u^{(k)}, \quad u_{n+1}^{(0)} = u_{n+\gamma} \tag{3.72}$$

and approximating

$$f_{n+1}^{(k+1)} = f_{n+1}^{(k)} + \left(\frac{\partial f}{\partial u}\right)_{n+1}^{(k)} \Delta u^{(k)} \tag{3.73}$$

where $k = 0, 1, 2, \ldots$ labels the Newton iterations. At each TR or BDF2 partial step, we iterate until the Newton method converges.

The Newton equation for the TR partial step is

$$\left[I - \gamma\frac{\Delta t}{2}\left(\frac{\partial f}{\partial u}\right)_{n+\gamma}^{(k)}\right]\Delta u^{(k)} = -\left(u_{n+\gamma}^{(k)} - u_n\right) + \gamma\frac{\Delta t}{2}\left(f_{n+\gamma}^{(k)} + f_n\right) \equiv -R_{\text{TR}} \tag{3.74}$$

where R_{TR} is the residual for (3.70).

The Newton equation for the BDF2 partial step is

---

[4] For BDF2, $e_l \approx -\frac{(1-\gamma)^2}{6(2-\gamma)}\Delta t^3 u'''$.

$$\left[I - \frac{1-\gamma}{2-\gamma}\Delta t \left(\frac{\partial f}{\partial u}\right)^{(k)}_{n+1} \right] \Delta u^{(k)}$$

$$= -\left(u^{(k)}_{n+1} - \frac{1}{\gamma(2-\gamma)} u_{n+\gamma} + \frac{(1-\gamma)^2}{\gamma(2-\gamma)} u_n \right) + \frac{1-\gamma}{2-\gamma}\Delta t f^{(k)}_{n+1} \equiv -R_{\text{BDF2}}$$
(3.75)

where R_{BDF2} is the residual for (3.71).

The timestep size Δt is adjusted dynamically within a window $[\Delta t_{\min}, \Delta t_{\max}]$ by monitoring a *divided-difference* estimate of the local error e_l:

$$e_l = k_\gamma \Delta t^3 u''' \approx 2 k_\gamma \Delta t \left(\frac{1}{\gamma} f_n - \frac{1}{\gamma(1-\gamma)} f_{n+\gamma} + \frac{1}{1-\gamma} f_{n+1} \right) \quad (3.76)$$

where

$$k_\gamma = \frac{-3\gamma^2 + 4\gamma - 2}{12(2-\gamma)}. \quad (3.77)$$

The three values of f employed in (3.76) have already been calculated in the most recent TRBDF2 timestep. The $||e_l||$ is minimized for $\gamma = 2 - \sqrt{2} \approx 0.59$.

Recall from (2.24) that a time integration method for $du/dt = au$, $\Re\{a\} < 0$, is *A-stable* if

$$\frac{||u_{n+1}||}{||u_n||} = ||G|| \leq 1 \quad (3.78)$$

for all $\Delta t > 0$; from (2.25), the time integration method is *L-stable* if it is A-stable and

$$\lim_{\Delta t \to \infty} \frac{||u_{n+1}||}{||u_n||} = \lim_{\Delta t \to \infty} ||G|| = 0. \quad (3.79)$$

For the heat/diffusion equation, TR is A-stable but not L-stable; backward Euler (first- and second-order) and TRBDF2 are L-stable.

The growth factor G for TRBDF2 for $du/dt = -\alpha u$, $\alpha > 0$, is given by ($\Delta = \gamma \alpha \Delta t > 0$):

$$G(\Delta t, \gamma) = \frac{\frac{1-\frac{\gamma\alpha\Delta t}{2}}{1+\frac{\gamma\alpha\Delta t}{2}} - (1-\gamma)^2}{\gamma(2-\gamma) + \gamma(1-\gamma)\alpha\Delta t} = \frac{\frac{2-\Delta}{2+\Delta} - (1-\gamma)^2}{\gamma(2-\gamma) + (1-\gamma)\Delta}. \quad (3.80)$$

For $\gamma = 1$, the BDF2 step disappears, and $G \to G_{TR}$; similarly, for $\gamma = 0$, the TR step disappears, the BDF2 step $\to$ TR, and $G \to G_{TR}$. In both cases, $k_\gamma \to -\frac{1}{12}$

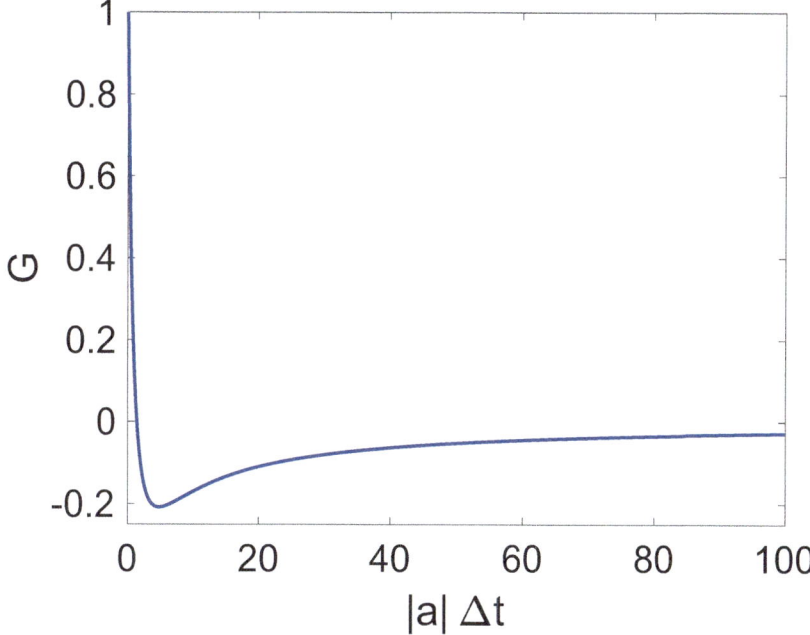

Fig. 3.13 TRBDF2 growth factor G as a function of $|a|\Delta t$ for $\gamma = 2 - \sqrt{2}$

(the TR value). Figure 3.13 illustrates G for $\gamma = 2 - \sqrt{2}$. Problem 3.18 derives G and proves that TRBDF2 is A-stable and L-stable.

Figure 3.1 for TRBDF2 with $\gamma = 2 - \sqrt{2}$ demonstrates graphically that TRBDF2 is Λ stable.

3.10 Equivalence Theorem for ODE IVPs

Theorem *Equivalence Theorem: For* consistent *numerical approximations,* stability *and* convergence *are equivalent.*

Lax proved the Equivalence Theorem for ODE initial value problems. The theorem applies as well to ODE boundary value problems, approximations to functions and integrals, and PDE initial value problems. Here we will prove the theorem for ODE initial value problems; the proof (due to Richtmyer) for PDE initial value problems is more difficult, since it involves functional analysis (see Sect. 5.3 on the Equivalence Theorem for PDEs).

We give two proofs, following Strang [35].

Approximate $Lu = f$ by $L_\Delta u_\Delta = f_\Delta$. *Consistency* implies $f_\Delta \to f$ and $L_\Delta u \to Lu$ as $n \to \infty$, $\Delta t \to 0$, with $n\Delta t = t$ fixed. *Stability* implies L_Δ^{-1}

remains uniformly bounded. Then as $n \to \infty$, $\Delta t \to 0$, with $n \Delta t = t$ fixed, u_Δ *converges* to u:

$$u - u_\Delta = L_\Delta^{-1}(L_\Delta u - Lu) + L_\Delta^{-1}(f - f_\Delta) \to 0. \qquad (3.81)$$

Let's look at the details for the initial value problem $du/dt = au$, $u(0) = u_0$. The exact solution is $u(t) = e^{at}u_0 = H^n u_0$, where $H = e^{a\Delta t}$ is the exact growth factor. The approximate solution is $u_n = G^n u_0$. *Stability* is equivalent to

$$|G^n| \le e^{Kn\Delta t} = e^{Kt} \qquad (3.82)$$

where K is a nonnegative constant independent of n. For $a > 0$, choose $K \ge a > 0$; for $a < 0$ and A-stability, set $K = 0$ with $e^{Kt} = 1$. Then both $|G|$ and $|H|$ are $\le e^{K\Delta t}$. *Consistency* implies

$$|G - H| \le C \Delta t^{p+1}, \quad p > 0. \qquad (3.83)$$

Convergence is equivalent to

$$G^n u_0 - H^n u_0 \to 0 \qquad (3.84)$$

as $n \to \infty$, $\Delta t \to 0$, with $n \Delta t = t$ fixed.

To show *stability* implies *convergence*, we use a telescoping identity:

$$\begin{aligned} G^n - H^n &= G^n - G^{n-1}H + G^{n-1}H - G^{n-2}H^2 + \ldots + GH^{n-1} - H^n \\ &= G^{n-1}(G-H) + G^{n-2}(G-H)H + \ldots + (G-H)H^{n-1}. \end{aligned} \qquad (3.85)$$

There is a factor of $G - H$ in every term, with $|G-H| \le C\Delta t^{p+1}$ by consistency. There is a nonnegative integer power of G in every term, which is bounded by stability. And there is a nonnegative integer power of H in every term, which is bounded, since the continuous problem is well-posed. There are $n = t/\Delta t$ terms in (3.85). Therefore as $n \to \infty$, $\Delta t \to 0$, with $n \Delta t = t$ fixed, we have convergence:

$$|G^n - H^n| \le \frac{t}{\Delta t} e^{Kt} C \Delta t^{p+1} = O(\Delta t^p) \to 0. \qquad (3.86)$$

Each step was necessary—and all steps together are sufficient—for convergence, so we have proved *stability* is equivalent to *convergence* for *consistent* numerical schemes.

Numerical error accumulates in a difference equation exactly via the telescoping series (3.85):

$$G^n - H^n = \sum_{j=1}^{n} G^{n-j}(G-H)H^{j-1}. \qquad (3.87)$$

H^{j-1} propagates the exact solution to time level $j-1$; $(G-H)$ is the local error going from time level $j-1$ to j; and G^{n-j} propagates this error forward with the difference method to time level n.

Problems

3.1 For $du/dt = f(u)$, prove the backward Euler method is first-order accurate, using the definition of the local truncation error LTE = $\Delta t\, \tau$. Be sure to calculate the constant multiplying Δt in the global error τ. *Hint:* Replace $f(u_{n+1}) = u'(t + \Delta t)$.

3.2 Prove backward Euler is A-stable and L-stable for the model problem $du/dt = au$ with $a < 0$.

3.3 The *modified midpoint rule (MMR)* for $du/dt = f(u)$ is

$$u_{n+1} = u_n + \Delta t\, f\left(\frac{1}{2}(u_n + u_{n+1})\right).$$

Note the subtle difference from the TR method (3.12). For the linear initial value problem $du/dt = au$, show MMR is second-order accurate and A-stable but not L-stable.

3.4

(a) Show the Adams-Bashforth-Moulton predictor-corrector method (3.19) and (3.20) is second-order accurate.
(b) Verify the method is stable for $u' = au$, $a < 0$, if $\Delta t \leq 2/|a|$.

3.5

(a) Show the fourth-order Adams-Bashforth-Moulton predictor-corrector formulas (3.21) and (3.22) are correct by fitting a cubic to the four most recent points in the derivative picture, extrapolating for (3.21), and then integrating under the cubic polynomials.
(b) Also find the stability restriction on Δt for $u' = au$, $a < 0$.

3.6 For $du/dt = au$, show RK2 is (a) second-order accurate and (b) conditionally stable. *Hints:* For (a), show the growth factor $G = 1 + a\Delta t + \frac{1}{2}a^2 \Delta t^2 \approx e^{a\Delta t}$ to second order in Δt; in (b), take $a < 0$, and require $-1 \leq G \leq 1$.

3.7 Show RK4 is fourth-order accurate for $du/dt = au$ by showing the growth factor $G = e^{a\Delta t}$ through fourth order in Δt. Also find the stability condition on Δt.

3.8 Run the program **ivp.m** that solves the initial value problem $dy/dt = -y$, $y(0) = 1$ using backward Euler and TR. First try ivp(10,2,1) to see a typical case with 10 steps, $t_f = 2$, and $\Delta t = 0.2$.

(a) What goes wrong at $\Delta t = 2$? Try ivp(10,20,1) with ten steps, $t_f = 20$, $\Delta t = 2$. *Hint:* Look at G_{BE} and G_{TR}. For $y' = ay$,

$$G_{BE} = \frac{1}{1 - a\Delta t}, \quad G_{TR} = \frac{1 + \frac{1}{2}a\Delta t}{1 - \frac{1}{2}a\Delta t}.$$

(b) What do the results for $\Delta t = 2.1$ illustrate? Try ivp(10,21,1) with ten steps, $t_f = 21$, $\Delta t = 2.1$. *Hint:* Look at G_{BE} and G_{TR}.

3.9 For the van der Pol oscillator in **rk2dyn.m** with $\epsilon = 4$, explore various initial conditions: for example, $w_0 = [1; 0]$, $w_0 = [1; -5]$, $w_0 = [-2; 5]$, etc. What can you conclude about the various initial conditions and the long time attractor? Discuss why this happens based on the competition (3.49) between positive friction (when $y^2 > \epsilon$) and negative friction (when $y^2 < \epsilon$).

3.10 Verify that our **rk2dyn.m** code for the nonlinear van der Pol oscillator (take $\epsilon = 4$) converges under mesh refinement by comparing and contrasting simulations with $\epsilon_R = 10^{-2}$, 10^{-3}, and 10^{-6} at $t = 50$ (keep $\epsilon_A = 10^{-9}$). Use RK2dyn([1;0],50,2).

3.11 Investigate the limits of predictability for the Lorenz equations using **lorenz2.m** by comparing the graphs of $y(t)$ with initial data $x(0) = 0$, $y(0) = 1$, $z(0) = 0$, and $y_\delta(t)$ with initial data $x(0) = 0$, $y(0) = 1 + \delta$, $z(0) = 0$, where $\delta = 10^{-6}$. Set $\sigma = 10$, $r = 28$, $b = 8/3$, $t_f = 40$, $\epsilon_R = 10^{-12}$, and $\epsilon_A = 10^{-15}$. Describe your results.

3.12 Find the largest δ to the nearest power of 10 (e.g., 10^{-15}), so that the plots of $y(t)$ and $y_\delta(t)$ agree to within a line width out to $t_f = 40$. Set $\sigma = 10$, $r = 28$, $b = 8/3$, $t_f = 40$, $\epsilon_R = 10^{-12}$, and $\epsilon_A = 10^{-15}$. Discuss your results.

3.13 Investigate the limits of predictability for the Lorenz equations using **lorenz3.m** by comparing the graphs of $y(t)$ computed with **ode45**, **ode23tb** (with analytical Jacobian), and **ode23tb** (with finite difference Jacobian). Set $\sigma = 10$, $r = 28$, $b = 8/3$, $t_f = 40$, $\epsilon_R = 10^{-12}$, and $\epsilon_A = 10^{-15}$. Take the initial data $x(0) = 0$, $y(0) = 1$, $z(0) = 0$. Describe your results.

3.14 Investigate fixed point and periodic orbits for the Lorenz equations by varying r with $\sigma = 10$ and $b = 8/3$. Try $r = 10$ (fixed point), $r = 350$ (simplest periodic orbit), and $r = 99.65$, 100.5, and 160 (complicated periodic orbits).

3.15 Show the Jacobian for the Lorenz equations is given by (3.62).

3.16 Apply **ode45** (explicit RK4/RK5) and **ode23tb** (implicit TRBDF2) to the stiff initial value problem (Moler [29], Sect. 7.9) governing the radius $y(t)$ of a spherical ball of flame:

$$y' = y^2 - y^3, \quad y(0) = \delta.$$

The flame propagation equation says that the radius of the ball of flame increases in proportion to its surface area (the $+y^2$ term) and decreases in proportion to its volume (the $-y^3$ term). Try

```
delta = 0.00001;
F = @(t,y) y^2 - y^3;
ode45(F,[0 1.2/delta],delta);
```

Then try the stiff solver

```
ode23tb(F,[0 1.2/delta],delta);
```

In each case, zoom in on $t = 1/\delta$ to see the local oscillations with the nonstiff solver **ode45**, and their cure with the stiff solver **ode23tb**.

3.17 For the initial value problem $du/dt = f(u)$, derive the leading local error term (including the constant) for TRBDF2 using the definition of the LTE:

$$e_l \approx \text{LTE} \approx k_\gamma \Delta t^3 u''', \quad k_\gamma = \frac{-3\gamma^2 + 4\gamma - 2}{12(2-\gamma)}.$$

Hint: Set $u_{n+\gamma} = u(t_{n+\gamma}) - e_l^{TR}$.

3.18 Derive the growth factor G for TRBDF2 for $du/dt = -\alpha u$, $\alpha > 0$, and show the method is A-stable and L-stable for $0 < \gamma < 1$. *Hints:* (i) First derive the growth factor (3.80). (ii) Then write the A-stability conditions as

$$-1 \leq G = \frac{\frac{2-\Delta}{2+\Delta} - \theta_1^2}{\theta_2 + \theta_1 \Delta} \leq 1$$

where $\Delta = \gamma \alpha \Delta t > 0$ and

$$0 < \theta_1 = 1 - \gamma < 1, \quad 0 < \theta_2 = \gamma(2-\gamma) < 1.$$

Note that

$$\theta_1^2 + \theta_2 = 1, \quad \theta_1 + \theta_2 - 1 = \gamma(1-\gamma) > 0.$$

(iii) Finally prove $-1 \leq G \leq 1$.

3.19

(a) Verify the TRBDF2 divided-difference estimate (3.76) of the local error e_l,
(b) Show $|k_\gamma|$ (and thus $||e_l||$) is minimized for $0 < \gamma = 2 - \sqrt{2} < 1$.

3.20 For $du/dt = f(u)$, show

$$||e_l|| = ||u(t_{n+1}) - u_{n+1}|| \approx \frac{3\gamma^2 - 4\gamma + 2}{3\gamma(1-\gamma)^2} ||u_1 - u_2||$$

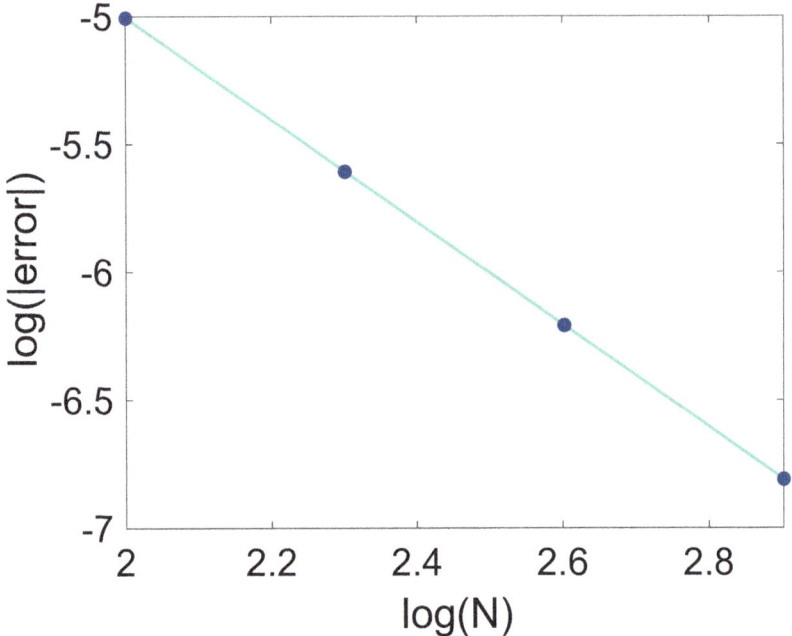

Fig. 3.14 Numerical errors $\log_{10}(\|y_{exact} - y_{comp}\|_1)$ vs. $\log_{10}(N)$ for TR using **ivp.m** for $dy/dt = -y$, $y(0) = 1$, with $N = 100, 200, 400$, and $800\Delta t$ at $t = 2$. The *straight line* is the theoretical value with log-log slope of -2

for the TR/TR version of calculating the local error for TRBDF2, where $u_1 \equiv u_{n+1} = u_{TRBDF2}$ and $u_2 \equiv u_{TR/TR}$ (see [23]).

3.21 In the TRBDF2 notes, verify (3.74) and (3.75) from the general Newton's formula for $du/dt = f(u)$

$$\frac{\partial R}{\partial u}\Delta u = -R$$

where R is the residual of the method (TR or BDF2).

3.22 Verify that our implementations of forward and backward Euler, TR, and TRBDF2 achieve the correct orders of accuracy Δt^p by comparing computed and exact solutions at $t = 2$ for the model problem $y' = -y$, $y(0) = 1$ on various temporal grids (Fig. 3.14). *Note:* The same exercise can be applied to the (damped) harmonic oscillator in **rk2.m** where an exact solution is available and to the nonlinear pendulum and van der Pol oscillator by letting a fine mesh solution stand in for the exact solution.

3.23 *Project:* Upgrade our RK2/RK2 solver to RK4/RK4, and test on the various problem types: damped harmonic oscillator, nonlinear pendulum, van der Pol oscillator, Shaw oscillator, Lorenz equations, and singularity detection.

Chapter 4
Numerical Methods for ODE BVPs

Two-point ODE boundary value problems[1] are ODEs of at least second order where boundary conditions for the solution are specified at the left and right boundaries. Along the x axis, allocate grid points

$$x_i = a + i\Delta x, \quad i = 0, 1, \ldots, N \qquad (4.1)$$

with fixed steps $\Delta x = (b-a)/N$. Boundary conditions will be imposed at $x_0 = a$ and $x_N = b$. A variable mesh spacing can be used, but the best grid refinement methods are multigrid (see [4], e.g.) and adaptive mesh refinement [3].

ODE boundary value problems are traditionally formulated for $y = y(x)$ on a spatial interval $[a, b]$, although equivalently we could consider $y = y(t)$ on a temporal interval $[t_a, t_b]$.

Discretized linear boundary value problems can be solved with a banded matrix solve (based on Gaussian elimination). For discretized nonlinear boundary value problems, Newton's method should be used to linearize the equations for the solution update, which then can be found with a banded matrix solve.

4.1 First and Second Derivative Matrices

First and second derivatives at the interior grid points $x_1, x_2, \ldots, x_{N-1}$ will be computed from solution values $y = [y_1, y_2, \ldots, y_{N-1}]$ and boundary conditions y_0 and y_N by

---

[1] It is possible to extend the analysis to ODE boundary value problems with boundary conditions at several points, but we will have no need to do so.

$$y' = \mathcal{D}^{(1)}y, \quad \mathcal{D}^{(1)} = \frac{1}{2h}\text{tridiag}[-1,\ 0,\ 1] \tag{4.2}$$

$$y'' = \mathcal{D}^{(2)}y, \quad \mathcal{D}^{(2)} = \frac{1}{h^2}\text{tridiag}[1,\ -2,\ 1] \tag{4.3}$$

where $h \equiv \Delta x$.

For $N = 10$ (boundary conditions are discussed below),

$$\mathcal{D}^{(1)} = \frac{1}{2h}\begin{bmatrix} 0 & 1 & 0 & 0 & 0 & 0 & 0 & 0 & 0 \\ -1 & 0 & 1 & 0 & 0 & 0 & 0 & 0 & 0 \\ 0 & -1 & 0 & 1 & 0 & 0 & 0 & 0 & 0 \\ 0 & 0 & -1 & 0 & 1 & 0 & 0 & 0 & 0 \\ 0 & 0 & 0 & -1 & 0 & 1 & 0 & 0 & 0 \\ 0 & 0 & 0 & 0 & -1 & 0 & 1 & 0 & 0 \\ 0 & 0 & 0 & 0 & 0 & -1 & 0 & 1 & 0 \\ 0 & 0 & 0 & 0 & 0 & 0 & -1 & 0 & 1 \\ 0 & 0 & 0 & 0 & 0 & 0 & 0 & -1 & 0 \end{bmatrix} \tag{4.4}$$

$$\mathcal{D}^{(2)} = \frac{1}{h^2}\begin{bmatrix} -2 & 1 & 0 & 0 & 0 & 0 & 0 & 0 & 0 \\ 1 & -2 & 1 & 0 & 0 & 0 & 0 & 0 & 0 \\ 0 & 1 & -2 & 1 & 0 & 0 & 0 & 0 & 0 \\ 0 & 0 & 1 & -2 & 1 & 0 & 0 & 0 & 0 \\ 0 & 0 & 0 & 1 & -2 & 1 & 0 & 0 & 0 \\ 0 & 0 & 0 & 0 & 1 & -2 & 1 & 0 & 0 \\ 0 & 0 & 0 & 0 & 0 & 1 & -2 & 1 & 0 \\ 0 & 0 & 0 & 0 & 0 & 0 & 1 & -2 & 1 \\ 0 & 0 & 0 & 0 & 0 & 0 & 0 & 1 & -2 \end{bmatrix}. \tag{4.5}$$

4.2 Linear Boundary Value Problems

As an example, let's discretize the linear boundary value problem

$$y'' + 2y' + y = 0, \quad y(0) = 1, \quad y(1) = 0. \tag{4.6}$$

We can compare our computed solution against the exact solution

$$y(x) = (1 - x)e^{-x}. \tag{4.7}$$

The discrete equations for $y = [y_1, y_2, \ldots, y_{N-1}]$ are

$$\mathcal{D}^{(2)}y + 2\mathcal{D}^{(1)}y + y + b \equiv Ay + b = 0 \tag{4.8}$$

4.3 Nonlinear BVPs and Newton's Method

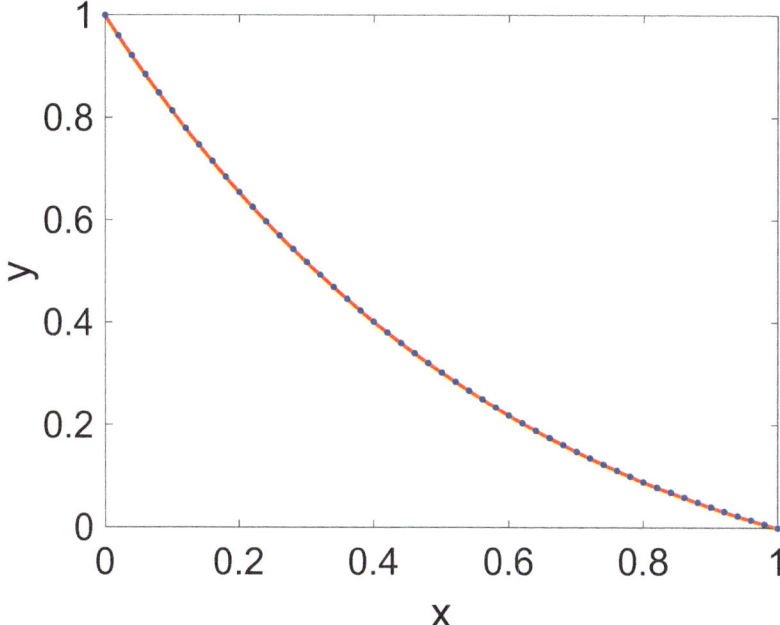

Fig. 4.1 Computed (*dots*) and exact (*curve*) solutions to the linear boundary value problem (4.6) using **bvp.m** with $50 \Delta x$

where

$$b = [y_0/h^2 - y_0/h, 0, \ldots, 0, y_N/h^2 + y_N/h] \qquad (4.9)$$

incorporates the coupling to the boundary conditions. The solution to the boundary value problem is given in Fig. 4.1 by numerically solving $Ay = -b$ using the backslash operator[2] in MATLAB: y = -A\b.

Figure 4.2 demonstrates that **bvp.m** achieves second-order accuracy by comparing computed and exact solutions on various spatial grids.

4.3 Nonlinear BVPs and Newton's Method

As a nonlinear example, let's compute the solution to the boundary layer equation [2]

---

[2] A tridiagonal solve in this case.

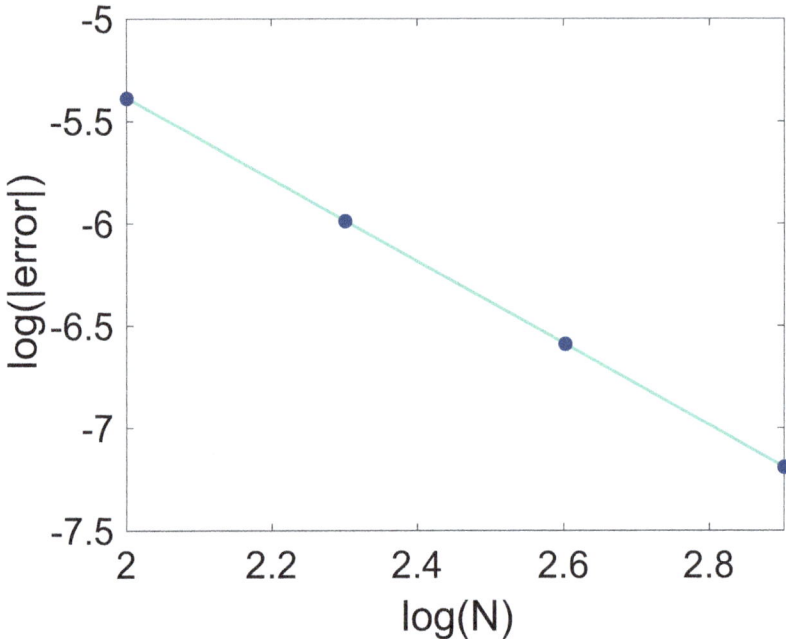

Fig. 4.2 Numerical errors $\log_{10}(||y_{exact} - y_{comp}||_1)$ vs. $\log_{10}(N)$ for the boundary value problem (4.6) using **bvp.m** with $N = 100, 200, 400$, and $800\Delta x$. The *straight line* is the theoretical value with log-log slope of -2

$$\epsilon y'' + 2y' + e^y = 0, \quad y(0) = 0, \quad y(1) = 0, \quad 0 < \epsilon \ll 1. \tag{4.10}$$

The discrete equations for $y = [y_1, y_2, \ldots, y_{N-1}]$ are

$$\epsilon \mathcal{D}^{(2)} y + 2\mathcal{D}^{(1)} y + e^y \equiv F(y) = 0. \tag{4.11}$$

To solve this system of nonlinear equations, we make a guess for y and then iterate using Newton's method until the norm of the residual $||F^{(k+1)}|| \le \epsilon_R ||F^{(0)}||$, $\epsilon_R = 10^{-6}$ (e.g.):

$$y^{(k+1)} = y^{(k)} + \Delta y, \quad y^{(0)} = 0, \quad k = 0, 1, 2, \ldots \tag{4.12}$$

$$0 = F\left(y^{(k+1)}\right) \approx F\left(y^{(k)}\right) + J \Delta y \tag{4.13}$$

where the Jacobian

$$J = \left.\frac{\partial F}{\partial y}\right|_{y^{(k)}} = \epsilon \mathcal{D}^{(2)} + 2\mathcal{D}^{(1)} + \text{diag}\left\{e^{y^{(k)}}\right\}. \tag{4.14}$$

4.3 Nonlinear BVPs and Newton's Method

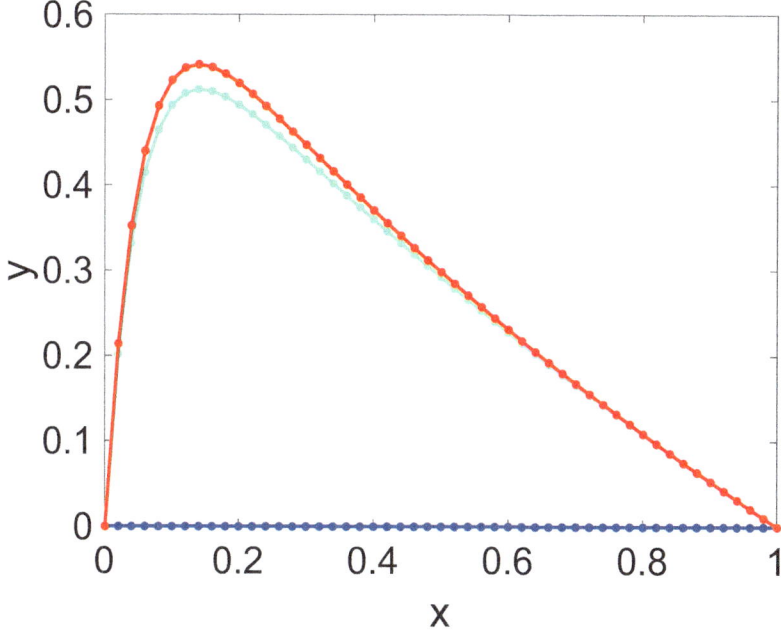

Fig. 4.3 Newton iterates 0 (*bottom dots* connected by a *curve*), 1 (*middle*), and 2 (*top*) from **layer.m** for the nonlinear boundary value problem (4.10) with $50\Delta x$ for $\epsilon = 0.1$

Solving for the corrections Δy, we get

$$J\Delta y = -F, \quad y \leftarrow y + \Delta y. \tag{4.15}$$

Newton iterates for $\epsilon = 0.1$ are shown in Fig. 4.3.

A uniform global approximation to $y(x)$ solving (4.10) is derived below for $\epsilon \to 0$:

$$y(x) = \ln\left(\frac{2}{x+1}\right) - \ln(2)\exp\left(\frac{-2x}{\epsilon}\right). \tag{4.16}$$

In this limit, the width of the boundary layer (location of the maximum of y) is given by

$$\mathcal{B} = -\frac{1}{2}\epsilon(\ln(\epsilon) - \ln(2\ln(2))). \tag{4.17}$$

Computed vs. approximate solutions are shown in Fig. 4.4, and computed vs. approximate boundary layer widths are shown in Fig. 4.5.

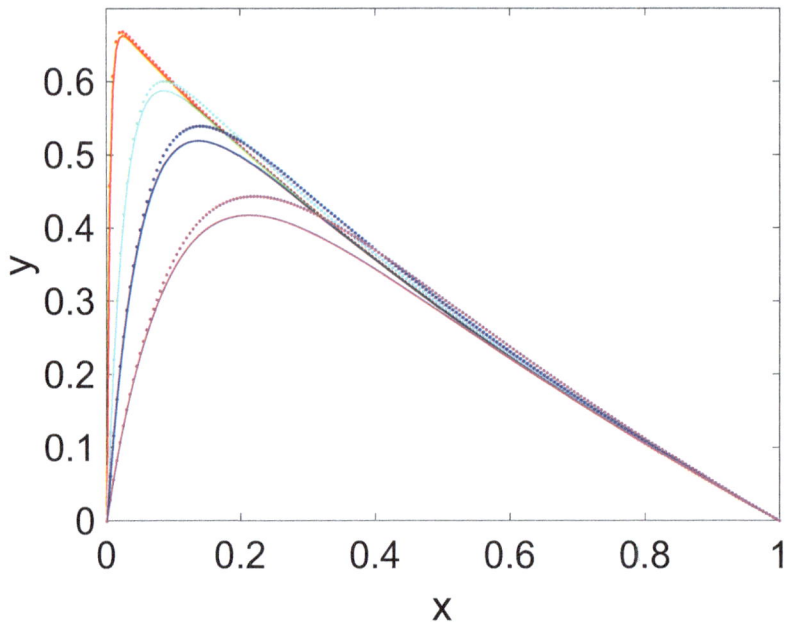

Fig. 4.4 *Dots* are the computed solutions from **layer.m** to the nonlinear boundary value problem (4.10) with $200\Delta x$ for $\epsilon = 0.01$ (top at the left), 0.05 (second from top), 0.1 (third from top), 0.2 (bottom); *curves* are the corresponding analytical approximations

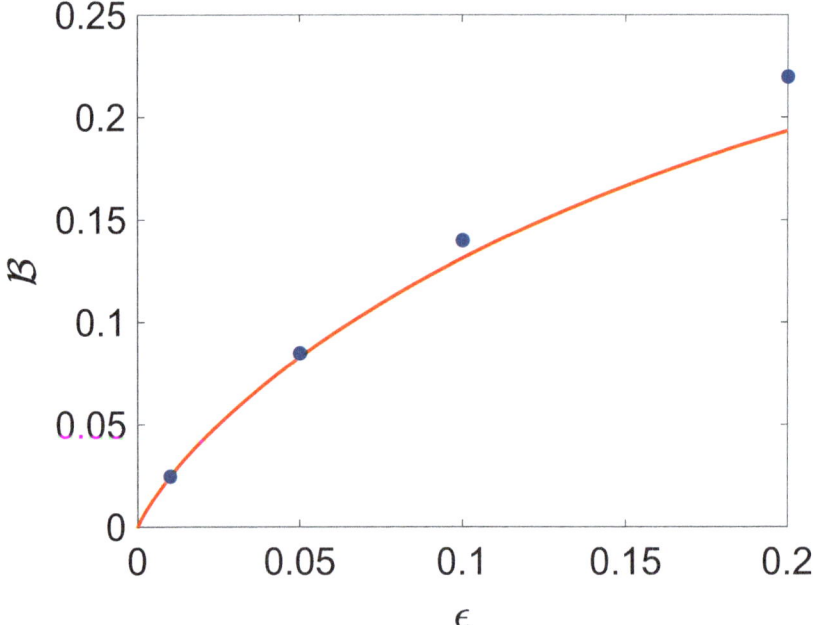

Fig. 4.5 Boundary layer widths $\mathcal{B}$ vs. ϵ from **layer.m** (*dots*) for the nonlinear boundary value problem (4.10) and analytical approximation (*curve*)

Derivation of the uniform global approximation

(i) In the outer laminar region, y is changing slowly, so we can neglect $\epsilon y''$ in comparison with the other two terms in (4.10):

$$2y'_{out} + \exp(y_{out}) = 0, \quad y_{out}(x) = \ln\left(\frac{2}{x+c}\right) = \ln\left(\frac{2}{x+1}\right) \quad (4.18)$$

to satisfy the boundary condition $y(1) = 0$.

(ii) In the inner boundary layer, y is changing rapidly. Set

$$y(x) = y_{in}(X) = y_{in}\left(\frac{x}{\epsilon}\right). \quad (4.19)$$

Then to leading order ϵ^{-1} in ϵ (noting $d/dx = \epsilon^{-1} d/dX$),

$$\frac{d^2 y_{in}}{dX^2} + 2\frac{dy_{in}}{dX} = 0 \quad (4.20)$$

$$y_{in}(X) = \alpha + \beta e^{-2X} = \alpha(1 - e^{-2X}) \quad (4.21)$$

to satisfy the boundary condition $y(0) = 0$. By requiring $y_{in}(X) = \ln(2)$ as $X \to \infty$ to match y_{out} as $x \to 0$, we obtain

$$y_{in}(X) = \ln(2)(1 - e^{-2X}). \quad (4.22)$$

(iii) *Asymptotic matching.* The uniform global approximation is

$$y(x) = \ln\left(\frac{2}{x+1}\right) - \ln(2)\exp\left(\frac{-2x}{\epsilon}\right) \quad (4.23)$$

since it agrees with y_{in} inside the boundary layer and with y_{out} in the laminar region.

Problems

4.1 Show **layer.m** converges under mesh refinement. Take $\epsilon = 0.05$, and superimpose the solutions in a figure for 25, 50, and 100 Δx.

4.2 (a) Discretize the residual for the nonlinear Poisson-Boltzmann boundary value problem

$$F(y) = y'' + \frac{2}{x}y' - e^y = 0, \quad y(1) = 1, \quad y(2) = 0.$$

Hint: $F_i = \sum_j \mathcal{D}_{ij}^{(2)} y_j + \ldots + b_i$, where i and $j = 1, 2, \ldots, N-1$ and where the boundary conditions are incorporated into

$$b = \left[\frac{1}{h^2} - \frac{2}{2h}, 0, \ldots, 0, 0\right].$$

(b) What is the discretized Jacobian $J_{ij} = \partial F_i/\partial y_j$? *Hint:* Your answer will involve $\mathcal{D}_{ij}^{(2)}$, $\mathcal{D}_{ij}^{(1)}$, and the Kronecker δ defined by

$$\delta_{ij} = \begin{cases} 1 \text{ if } i = j \\ 0 \text{ if } i \neq j \end{cases}$$

4.3 Develop a program **nonlinPB.m** to solve the nonlinear boundary value problem in Problem 4.2.

4.4 Cf. Moler [29], 7.22 (d). In solving the nonlinear boundary value problem $y'' = y^2 - 1$, $y(0) = 0$, $y(1) = 1$ using Newton's method, we need to discretize the residual

$$F = y'' - y^2 + 1$$

and the Jacobian $J = \partial F/\partial y$.

(a) What is the discretized residual

$$F_i = \sum_j \mathcal{D}_{ij}^{(2)} y_j + \ldots + b_i$$

where i and $j = 1, 2, \ldots, N-1$ and where the boundary conditions are incorporated into b_i? *Hint:* $b = [0, 0, \ldots, 0, 1/h^2]$.

(b) What is the discretized Jacobian $J_{ij} = \partial F_i/\partial y_j$? *Hint:* Your answer will involve both $\mathcal{D}_{ij}^{(2)}$ and the Kronecker δ_{ij}.

4.5 Develop a program **nonlinBVP.m** to solve the nonlinear boundary value problem in Problem 4.4.

4.6 If an exact solution is not available, we can still estimate the order of accuracy of a method by using a fine grid solution y_{fine} to stand in for the exact solution. Verify that our implementation of the nonlinear layer boundary value problem (4.10) achieves second-order accuracy by comparing computed solutions (take $\epsilon = 0.1$) on spatial grids with $N = 100, 200, 400$, and $800\Delta x$ with a fine grid solution with $1600\Delta x$. *Hint:* First compute y_{fine} in **layer.m** using $1600\Delta x$ and then project onto a grid with $N\Delta x$.

Chapter 5
Overview of PDEs

Second-order linear PDEs in two variables with real coefficients (all the PDEs examined in this book will have real coefficients except Schrödinger's equation) can be classified as parabolic, elliptic, or hyperbolic. It is possible to extend this classification to the modes of systems of linear and nonlinear PDEs in two or more variables. Numerical methods that work for linear parabolic, elliptic, or hyperbolic PDEs in two variables can be expected to work, when properly generalized, for the parabolic, elliptic, or hyperbolic modes, respectively, of systems of nonlinear PDEs in two or more variables.

A generic PDE system will likely have modes that do not fall into any of these categories, but most of the PDEs of science have modes that do. And with judiciousness, frequently methods developed for parabolic, elliptic, or hyperbolic problems can be extended to apply to PDEs like Schrödinger's equation (imaginary coefficient $i\hbar$ in front of ∂_t) or the Korteweg-De Vries (KdV) equation ($u_t + u u_x = -u_{xxx}$) that do not fall into these categories.

Intuitively, solutions to a hyperbolic PDE (like the wave equation) or PDE system (like gas dynamics) propagate as waves with finite velocity (Figs. 5.1, 1.1, and 1.2), while solutions to a parabolic PDE (like the heat/diffusion equation) "ooze" outward in time (Figs. 5.2 and 6.6). An elliptic PDE can be viewed as the steady-state limit of a corresponding parabolic PDE (see (5.3) and (5.4) below for an example), although this approach is usually not recommended for computations, since it is much less efficient than directly solving the time-independent problem.

As mentioned in Chap. 1, there are three differential equation classes with related numerical methods: (i) ODE initial value problems and parabolic PDEs, (ii) ODE boundary value problems and elliptic PDEs, and (iii) hyperbolic PDEs. The first and third classes involve evolution in time, while the second class is time-independent.

A proof of the Equivalence Theorem for PDE initial value problems is given at the end of the chapter.

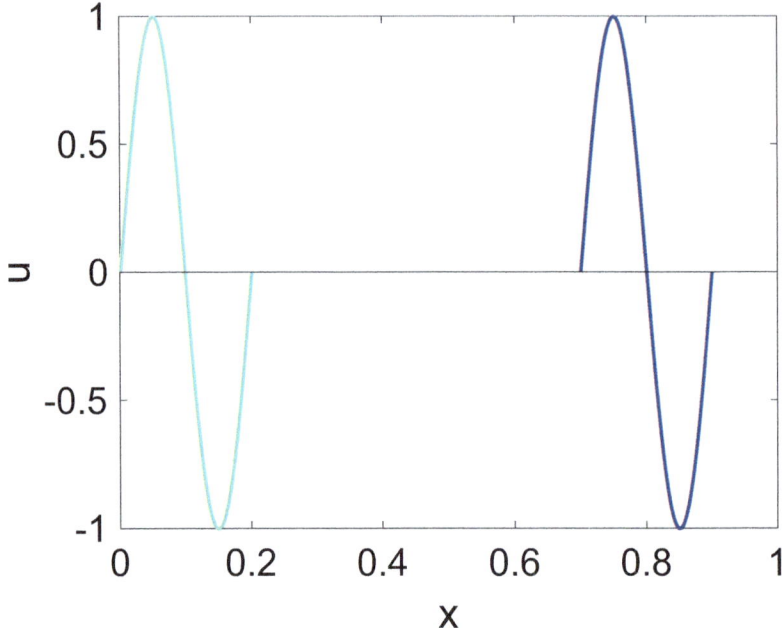

Fig. 5.1 Initial conditions (left) and upwind solution (right) at $t = 0.7$ to the first-order wave equation computed with **wave1.m** with $100\Delta x$ and $\Delta t = \Delta x/c$, illustrating finite propagation velocity ($c = 1$) of the wave. In general, hyperbolic waves can change shape as they propagate (as in the Frontispiece)

5.1 Classifying Linear PDEs

Most of the PDEs of science can be classified as hyperbolic, parabolic, elliptic, or Schrödinger (related to parabolic)—or as a mixture of modes of these types.

In particular, the general second-order linear PDE in two variables with real coefficients

$$au_{tt} + 2bu_{tx} + cu_{xx} + du_t + eu_x + fu = g \qquad (5.1)$$

can always be transformed into the wave equation, the heat/diffusion equation, or Poisson's equation (see Table 5.1).

Dirichlet boundary conditions specify $u(\mathbf{x}_\mathcal{B}, t)$ on the boundary $\mathbf{x}_\mathcal{B}$, Neumann boundary conditions specify $\hat{\mathbf{n}} \cdot \nabla u(\mathbf{x}_\mathcal{B}, t)$ on the boundary, where $\hat{\mathbf{n}}$ is a unit outward-pointing normal vector to the boundary, and Cauchy boundary conditions are determined from how information flows along *characteristics* in hyperbolic

5.1 Classifying Linear PDEs

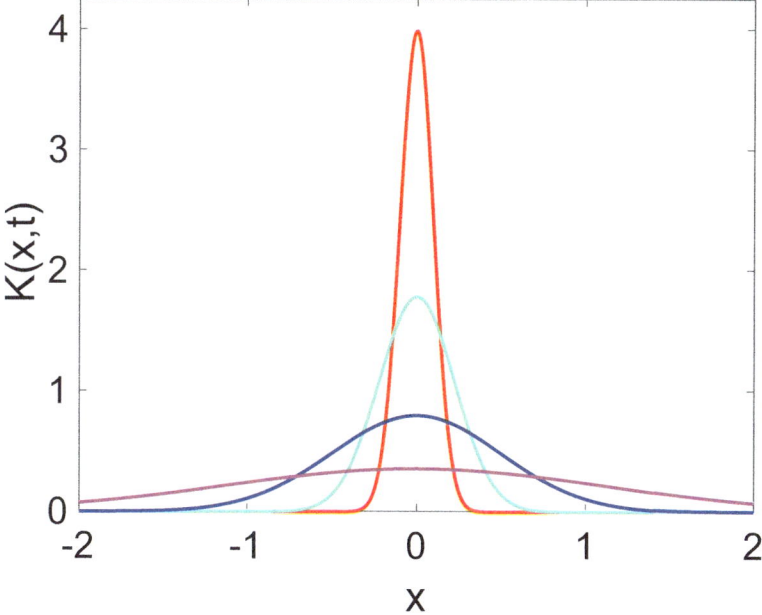

Fig. 5.2 Analytical Gaussian solutions $K(x, t)$ to the linear heat/diffusion equation, from top peak to bottom peak as time progresses. The area under each Gaussian is one

Table 5.1 Second-order PDE types. For the heat/diffusion equation case $\kappa > 0$ (if $\kappa < 0$, we have the ill-posed backward heat/diffusion equation). Note that Poisson's equation is usually written as $u_{xx} + u_{yy} = \cdots$

$b^2 - ac$	Transforms to	Equation	Type	Boundary conditions
+ve	$u_{tt} - \alpha^2 u_{xx} = \cdots$	Wave	*hyperbolic*	Cauchy
0	$u_t = \kappa u_{xx} + \cdots$	Heat/diffusion	*parabolic*	Dirichlet/Neumann
−ve	$u_{tt} + u_{xx} = \cdots$	Poisson	*elliptic*	Dirichlet/Neumann

PDEs (see Chap. 8). For parabolic and elliptic PDEs, Dirichlet or Neumann (or Robin[1]) boundary conditions can be specified on different parts of the boundary.

This classification can be extended into higher dimensions and also applied to (modes of) linear and nonlinear PDE systems. However, modes of a random PDE equation or system in general will not fall into any of these classes. The *dispersive* KdV equation is a well-known example: $u_t + u u_x = -u_{xxx}$.

In Table 5.2, the first- and second-order wave equations, Maxwell's equations, the inviscid Burgers equation, and the Euler equations of gas dynamics (a nonlinear system of PDEs) are hyperbolic. The linear and nonlinear diffusion equations, the advection-diffusion equation, and the viscous Burgers equation are parabolic.

---

[1] Robin boundary conditions specify a linear combination of Dirichlet and Neumann conditions.

Table 5.2 Paradigmatic examples of hyperbolic, parabolic, and elliptic PDEs

Hyperbolic	Parabolic	Elliptic
Wave equations	Heat/diffusion equation	Poisson's equation
Maxwell's equations	Advection-diffusion equation	Laplace's equation
Inviscid Burgers equation	Viscous Burgers equation	
Gas dynamics	Navier-Stokes equations	

The incompressible Navier-Stokes equations are called incompletely parabolic (parabolic transport equations plus an elliptic constraint $\nabla \cdot \mathbf{v} = 0$).

With a factor of i in front of the time derivative, the heat/diffusion equation becomes the Schrödinger equation

$$i\hbar \frac{\partial u}{\partial t} = -\frac{\hbar^2}{2m} \frac{\partial^2 u}{\partial x^2} + \cdots \tag{5.2}$$

which is related to parabolic PDEs but can be thought of as a separate type.

Elliptic and parabolic PDEs have the same boundary condition types: Dirichlet or Neumann or Robin. An elliptic PDE can be viewed as the steady-state ($u_t \to 0$ as $t \to \infty$) limit of a corresponding parabolic PDE. For example, the steady-state solution $u(\mathbf{x}, t \to \infty)$ of the nonlinear parabolic PDE

$$\frac{\partial u}{\partial t} = \nabla \cdot (\varepsilon(u) \nabla u) + \rho(\mathbf{x}) \tag{5.3}$$

is the solution $\tilde{u}(\mathbf{x})$ of the nonlinear elliptic PDE

$$\nabla \cdot (\varepsilon(\tilde{u}) \nabla \tilde{u}) = -\rho(\mathbf{x}). \tag{5.4}$$

5.2 Classifying General Nonlinear Systems of PDEs

The nonlinear system of PDEs (note that higher-order time derivatives can always be replaced by adding more state variables and equations)

$$w_t + F(w, w_x, w_{xx}, \ldots) = 0 \tag{5.5}$$

is classified by expanding w as a Fourier perturbation about a constant nonlinear solution $\overline{w}$

$$w = \overline{w} + e^{-\sigma t + ikx} \delta w \tag{5.6}$$

where σ is the decay rate and k is the wave number of the perturbation, and then linearizing with respect to δw:

5.2 Classifying General Nonlinear Systems of PDEs

$$\sigma \delta w = S \delta w \tag{5.7}$$

where S is the *symbol* of the linearized PDE system.

The asymptotic eigenvalues σ of S as $k \to \infty$ (we are interested in the *local* behavior of the perturbation) determine the mathematical type of the various modes of (5.5):

$$\sigma \sim \begin{cases} \pm ik & \text{hyperbolic} \\ k^2 & \text{parabolic} \\ -k^2 & \text{unstable} \\ \pm ik^2 & \text{Schrodinger} \\ \pm ik^3 & \text{dispersive} \\ k^4 & \text{fourth-order diffusive} \\ -k^4 & \text{unstable} \end{cases} \tag{5.8}$$

Typically elliptic modes decouple as a subproblem in S as in Poisson's equation $\nabla^2 \phi = -\rho$ in electro-gas dynamics or the elliptic constraint in the Navier-Stokes equations $\nabla \cdot \mathbf{u} = 0$.

To see in detail how the different contributions to σ affect the advection, growth, decay, or oscillation of the perturbation, consider the linear scalar PDE

$$u_t \pm cu_x = Du_{xx} + i\hbar u_{xx} \pm \beta u_{xxx} - \epsilon u_{xxxx}. \tag{5.9}$$

Recall a general function can represented by a Fourier integral

$$u(x,t) = \int_{-\infty}^{\infty} \frac{dk}{2\pi} e^{ikx} u_k(t). \tag{5.10}$$

For each wave number k, the solution behaves for small t like

$$e^{ikx} u_k(t) \sim e^{-\sigma t + ikx} = \exp\left\{ ik(x \mp ct) - Dk^2 t - i\hbar k^2 t \mp i\beta k^3 t - \epsilon k^4 t \right\} \tag{5.11}$$

where the coefficients D (diffusive) and ϵ (fourth-order diffusive) must be ≥ 0 for stability[2] (note the minus sign in front of ϵ on the right-hand side of (5.9)), while the advective c, Schrödinger[3] $i\hbar$, and dispersive β coefficients do not play a role in stability. The solution (5.11) consists of a traveling wave term (if $c \neq 0$), plus diffusive decay (for $D > 0$), Schrödinger oscillation (for $\hbar \neq 0$), dispersive oscillation (for $\beta \neq 0$), and fourth-order diffusive decay (for $\epsilon > 0$) terms. The

---

[2] It is possible to stabilize a negative D with a positive ϵ.
[3] We have set $2m = 1$ here.

mathematical type of the PDE (5.9) is determined by the dominant term in (5.11) as $k \to \infty$.

5.3 Equivalence Theorem for PDE IVPs

The proof of the Equivalence Theorem for linear PDE initial value problems was originally given by Richtmyer. Our proof is based on Gottlieb and Orszag [18]. Discretize the linear PDE initial value problem

$$\frac{\partial u}{\partial t} = Lu + b, \quad u(x, t = 0) = u(0) \tag{5.12}$$

in space, where L is the spatial derivative operator ($\alpha \partial_x$ or $\beta \nabla^2$, e.g.) and $u \in \mathcal{H}$, with $\mathcal{H}$ a Hilbert space,[4] to obtain the semi-discrete problem

$$\frac{\partial}{\partial t} u_N(x, t) = L_N u_N(x, t) + b_N(x, t). \tag{5.13}$$

(i) We will first show stability[5]

$$||e^{L_N t}|| \leq K(t) \tag{5.14}$$

for all N, where $K(t)$ is a bounded function of t, implies convergence. Note that

$$\frac{\partial}{\partial t}(u - u_N) = Lu - L_N u_N + b - b_N = L_N(u - u_N) + Lu - L_N u + b - b_N \tag{5.15}$$

has the formal solution (Duhamel's principle)

$$u(t) - u_N(t) = e^{L_N t}[u(0) - u_N(0)]$$
$$+ \int_0^t e^{L_N(t-s)}[Lu(s) - L_N u(s) + b(s) - b_N(s)] ds. \tag{5.16}$$

Taking the norm of both sides and repeatedly using the triangle inequality, we obtain

---

[4] A complete inner product space.
[5] The operator norm is defined below in the Principle of Uniform Boundedness theorem. For A-stability, $K(t) = 1$.

$$\|u(t) - u_N(t)\| \le K(t) \|u(0) - u_N(0)\|$$

$$+ \int_0^t K(t-s) [\, \|Lu(s) - L_N u(s)\| + \|b(s) - b_N(s)\| \,] ds \to 0 \quad (5.17)$$

as $N \to \infty$ since K is bounded by stability and the three terms on the right-hand side inside $\|\cdot\| \to 0$ by consistency.

(ii) Conversely, to show convergence implies stability, we use the reverse triangle inequality:

$$0 \le \left| \|e^{L_N t} u\| - \|e^{L t} u\| \right| \le \|e^{L_N t} u - e^{L t} u\| \to 0 \quad (5.18)$$

as $N \to \infty$ by convergence. Since $\|e^{Lt} u\|$ is bounded by well-posedness, $\|e^{L_N t} u\|$ is bounded.

However, $\|e^{L_N t} u\|$ may depend on u and t. We make use of the following:

Theorem *Principle of Uniform Boundedness (Richtmyer and Morton): If $\|e^{L_N t} u\|$ is bounded as $N \to \infty$ for each t and for each $u \in \mathcal{H}$, then*

$$\|e^{L_N t}\| = \max_{u \in \mathcal{H}} \frac{\|e^{L_N t} u\|}{\|u\|} \quad (5.19)$$

is bounded as $N \to \infty$ for each t.

Problems

5.1 Use the Fourier method of Sect. 5.2 and linearization to describe the behavior of the KdV equation $u_t + u u_x = -u_{xxx}$ as the wave number $k \to \infty$.

5.2 Use the methods of Sect. 5.2 to classify the behavior of the Navier-Stokes equations

$$\frac{\partial \mathbf{u}}{\partial t} + \mathbf{u} \cdot \nabla \mathbf{u} = -\frac{1}{\rho} \nabla P + \nu \nabla^2 \mathbf{u}$$

$$\nabla \cdot \mathbf{u} = 0$$

where P is the pressure, as the wave number $k \to \infty$. *Hint:* Calculate the eigenvalues σ of the symbol $\mathcal{S}$ of the linearized PDE system.

5.3 Show the formal solution (Duhamel's principle) (5.16) does in fact satisfy the PDE (5.15) with the appropriate initial conditions.

Chapter 6
Numerical Methods for Parabolic PDEs

The heat/diffusion equation is our primary example of a *parabolic* PDE (Laplace's equation is *elliptic* and the wave equation is *hyperbolic*).

The diffusion equation and the heat equation are identical. In 1D, the linear diffusion equation is

$$\frac{\partial u}{\partial t} = D \frac{\partial^2 u}{\partial x^2}, \quad u(x, t = 0) = u_0(x) \tag{6.1}$$

where $u(x, t)$ is the number density of diffusing particles and $D > 0$ is the (constant) diffusion coefficient. In applied mathematics, we often set $D = 1$ for simplicity (D can be absorbed into the timescale in ∂t).

Mathematically appropriate boundary conditions for parabolic PDEs are Dirichlet ($u(x_\mathcal{B}, t)$ is specified on the boundary $x_\mathcal{B}$) or Neumann (in 1D, $u_x(x_\mathcal{B}, t)$ is specified on the boundary) or Robin (a linear combination of Dirichlet and Neumann conditions is specified on the boundary). Dirichlet or Neumann or Robin boundary conditions can be specified on different parts of the boundary.

In 3D, the linear diffusion equation is

$$\frac{\partial u}{\partial t} = D \nabla^2 u, \quad u(\mathbf{x}, t = 0) = u_0(\mathbf{x}). \tag{6.2}$$

For the heat equation, $u(\mathbf{x}, t) = T - T_a$ is the temperature with respect to a reference temperature T_a, which can conveniently be taken to be the ambient temperature, and D is replaced by the thermal conductivity $\kappa > 0$:

$$\frac{\partial u}{\partial t} = \kappa \nabla^2 u, \quad u(\mathbf{x}, t = 0) = u_0(\mathbf{x}). \tag{6.3}$$

The heat equation describes the diffusion of heat energy, since the thermal energy density Q_{heat} is proportional to the temperature T: $\Delta Q_{heat} = C \Delta T$, where C is the specific heat capacity of the substance at constant pressure or volume.

If the diffusion coefficient $D = D(\mathbf{x}, u)$ depends on $\mathbf{x}$ and/or u, the diffusion equation in 3D becomes

$$\frac{\partial u}{\partial t} = \nabla \cdot (D(\mathbf{x}, u) \nabla u). \tag{6.4}$$

In 3D, Neumann boundary conditions specify the normal derivative $\hat{\mathbf{n}} \cdot \nabla u(\mathbf{x}_\mathcal{B}, t)$, where $\hat{\mathbf{n}}$ is a unit outward-pointing normal vector to the boundary. Physically speaking, a Dirichlet boundary condition fixes the concentration (or temperature) u at the boundary ("coupling to an ambient bath"), while a Neumann boundary condition specifies the particle (or heat) flux

$$-(D \nabla u) \cdot \hat{\mathbf{n}} \tag{6.5}$$

through the boundary. In particular, a homogeneous Neumann boundary condition imposes zero flux through the boundary.

The nonlinear diffusion equation governs many important physical applications like semiconductor process modeling. In applying an implicit method like TRBDF2 to the nonlinear diffusion equation, we will use Newton's method to linearize the discrete equations.

Henceforth we will generally refer to the heat/diffusion equation simply as the diffusion equation.

Both the linear and nonlinear diffusion equations are examples of parabolic conservation laws, where $Q = \int u(x,t)\,dx$ (in 1D) or $Q = \int u(\mathbf{x},t)\,dV$ (in 3D) is conserved. We will always use conservative numerical methods for conservation laws—all of our numerical methods for the diffusion equation are conservative by design.

The diffusion equation is *stiff*, i.e., solutions generally have two or more widely disparate timescales. The stiffness is apparent from two points of view:

(i) The exact diffusion equation initial value problem is infinitely stiff. For bounded spatial intervals, the disparate timescales can be seen in the Fourier solution (6.20) (cf. Fig. 6.1), which has timescales $\bar{t} = \bar{l}^2/(Dn^2)$ for $n = 1, 2, \ldots$, where $\bar{l}$ is a typical length scale in the problem ($\bar{l} \sim \pi$ for the Fourier solution (6.20)).
(ii) The discretized diffusion equation initial value problem is stiff, but not infinitely stiff. Discretizing the diffusion equation in space on the unit interval with second-order accurate central differences yields $u_t = D\mathcal{D}^{(2)}u$, where $\mathcal{D}^{(2)} = $ tridiag$[1, -2, 1]/h^2$ is the $n \times n$ second derivative matrix with n interior grid points plus Dirichlet boundary conditions ($h = 1/(n+1)$ here). The eigenvalues of $D\mathcal{D}^{(2)}$ are

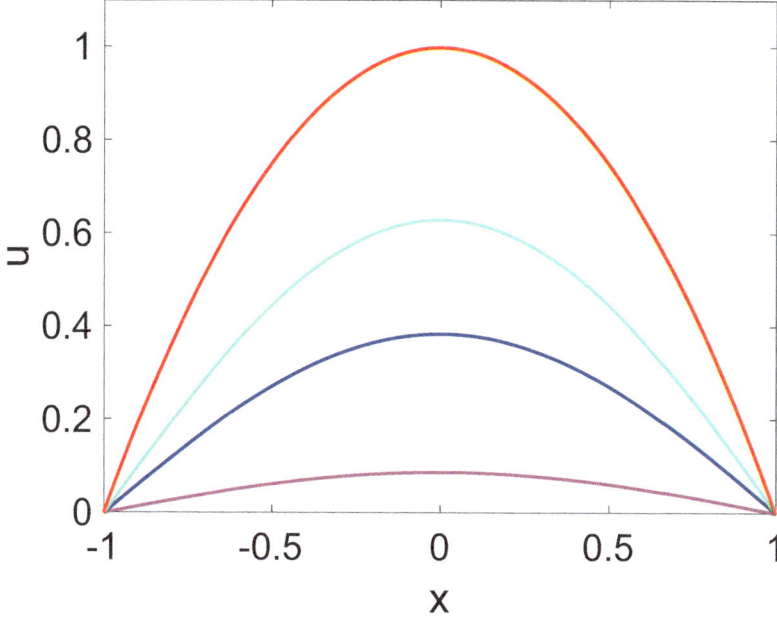

Fig. 6.1 Initial conditions $u_0(x) = 1 - x^2$ and computed solutions using **diffusion2.m** (TRBDF2) to the linear diffusion equation on a bounded spatial domain with $100 \Delta x$ and homogeneous Dirichlet boundary conditions: from top to bottom as time progresses

$$\lambda_j = \frac{2D}{h^2}(\cos(j\pi h) - 1), \quad j = 1, 2, \ldots, n \qquad (6.6)$$

(see Problem 6.1). These eigenvalues all lie on the negative real axis, with $\lambda_n \approx -4D/h^2$ the largest eigenvalue in magnitude. For the forward Euler method, $|1 + \lambda_n \Delta t|$ must be ≤ 1 or $\Delta t \leq 2h^2/D$ for stability, which is a manifestation of stiffness: Even when the computed solution is smooth and we would like to take large timesteps with $\Delta t \sim h$, we are restricted by stability to $\Delta t \sim h^2$. Similar remarks apply to all explicit methods for parabolic PDEs. Therefore for these PDEs, we always use a method like TRBDF2, which is both A-stable (to avoid timestep restrictions) *and* L-stable (to avoid local oscillations).

Our standard timestepping methods for ODE initial value problems all work for the diffusion equation (with second-order accurate central differences for spatial derivatives): forward Euler, backward Euler, TR, and TRBDF2. Table 6.1 lists the numerical methods we will discuss in this chapter. Since TRBDF2 is second-order accurate and L-stable, it is the method of choice for linear and nonlinear parabolic PDEs.

We will always dynamically adjust the timestep for parabolic and hyperbolic PDEs. For parabolic PDEs, Δt is based on an estimate of the local error, by

Table 6.1 Numerical methods for parabolic PDEs. Note that the trapezoidal rule (TR) method for the diffusion equation is also known as the Crank-Nicolson method

Method	Type	Order	Stability
Forward Euler	Explicit	First	$\Delta t \leq \Delta x^2/(2D)$
Backward Euler	Implicit	First	L-stable
TR	Implicit	Second	A-stable
TRBDF2	Implicit	Second	L-stable

comparing two computed solutions at the new time level, or from a divided-difference formula in TRBDF2. For hyperbolic PDEs, Δt is adjusted dynamically based on the CFL condition:

$$\Delta t \leq \frac{\Delta x}{\max\{|\lambda|\}} \tag{6.7}$$

where $\max\{|\lambda|\}$ is the maximum characteristic speed in magnitude on the grid at the beginning of the timestep. Again, for some hyperbolic schemes, there may be a fractional constant in the CFL condition, for example, $\Delta t \leq \frac{1}{2}\Delta x/\max\{|\lambda|\}$.

In discussing numerical methods, the exact solution $u(x_i, t_n)$ is approximated by u_i^n.

For another general introduction to solving parabolic PDEs, see LeVeque's *Finite Difference Methods for Ordinary and Partial Differential Equations* [26].

6.1 Heat/Diffusion Equation

The *diffusion* ($\equiv$ heat) equation is ($D > 0$)

$$\frac{\partial u}{\partial t} = D\frac{\partial^2 u}{\partial x^2}, \quad u(x, t = 0) = u_0(x). \tag{6.8}$$

The *fundamental solution* or *kernel* $K(x, t)$ of the diffusion equation (Fig. 5.2) is

$$K(x, t) = \frac{1}{\sqrt{4\pi Dt}} \exp\left\{-\frac{x^2}{4Dt}\right\} \tag{6.9}$$

which satisfies the initial value problem

$$\frac{\partial K}{\partial t} = D\frac{\partial^2 K}{\partial x^2}, \quad K(x, t = 0) = \delta(x) \tag{6.10}$$

where $\delta(x)$ is the Dirac δ function. Note that $\int_{-\infty}^{\infty} K(x, t)\, dx = 1$ for any t. $K(x - y, t)$ is also the Green's function $G(x, y; t)$ for the homogeneous diffusion equation.

6.2 Fourier Solution to the Diffusion Equation

The Dirac δ function is defined by

$$\delta(x) = \begin{cases} 0 & x \neq 0 \\ \infty & x = 0 \end{cases} \tag{6.11}$$

with $\int_{-\infty}^{\infty} \delta(x)\,dx = 1$ and $\int_{-\infty}^{\infty} \delta(x-a)f(x)\,dx = f(a)$. The δ function can be thought of as the limit of a sequence of Gaussians as the width $\to 0$ and the height $\to \infty$:

$$\delta(x) = \lim_{t \to 0} K(x,t) = \lim_{t \to 0} \frac{1}{\sqrt{4\pi Dt}} \exp\left\{-\frac{x^2}{4Dt}\right\}. \tag{6.12}$$

Note that since $\int_{-\infty}^{\infty} K(x,t)\,dx = 1$ for any t, in particular $\int_{-\infty}^{\infty} \delta(x)\,dx = 1$.
The *general solution* to the diffusion equation is

$$u(x,t) = \int_{-\infty}^{\infty} K(x-y,t)\,u_0(y)\,dy = \frac{1}{\sqrt{4\pi Dt}}\int_{-\infty}^{\infty} \exp\left\{-\frac{(x-y)^2}{4Dt}\right\}u_0(y)\,dy. \tag{6.13}$$

Formally, the propagation speed in the general solution is infinite, since some small effects of the initial conditions propagate immediately to $x = \pm\infty$ for any $t > 0$; however, it is more realistic to consider a physical propagation speed of the bulk of the diffusing substance based on the width $\sigma = \sqrt{2Dt}$ of the Gaussian in (6.13):

$$v_{parab} = \frac{d\sigma}{dt} = \sqrt{\frac{D}{2t}} \tag{6.14}$$

for $t > 0$.

The backward diffusion equation $u_t = -Du_{xx}$ ($D > 0$) obtained by $t \to -t$ is unstable for any $t > 0$, since now both $K \sim \exp\{+x^2/(4Dt)\}$ and u blow up as $x \to \pm\infty$.

6.2 Fourier Solution to the Diffusion Equation

The *Fourier solution* to the initial/boundary value problem for the diffusion equation with homogeneous Dirichlet boundary conditions (in this section D is set to 1)

$$u_t = u_{xx}, \quad u(0,t) = 0 = u(\pi,t) \text{ for } t > 0, \quad u(x,t=0) = u_0(x) \tag{6.15}$$

is obtained by making a Fourier sine expansion (since it automatically satisfies the boundary conditions)

$$u(x,t) = \sum_{n=1}^{\infty} a_n(t) \sin(nx) \tag{6.16}$$

where

$$a_n(0) = \frac{2}{\pi} \int_0^{\pi} \sin(nx) u_0(x)\, dx. \tag{6.17}$$

Plugging the Fourier expansion into the diffusion equation and using orthogonality of the sines (see Appendix A.4), we have

$$\sum_{n=1}^{\infty} \frac{da_n}{dt} \sin(nx) = -\sum_{n=1}^{\infty} n^2 a_n \sin(nx) \tag{6.18}$$

$$\frac{da_n}{dt} = -n^2 a_n, \quad a_n(t) = a_n(0) e^{-n^2 t} \tag{6.19}$$

and therefore

$$u(x,t) = \sum_{n=1}^{\infty} a_n(0) e^{-n^2 t} \sin(nx). \tag{6.20}$$

Note again that the backward diffusion equation $u_t = -u_{xx}$ obtained by $t \to -t$ is unstable for any $t > 0$, since $e^{n^2 t} \to \infty$ as $n \to \infty$.

The Fourier solution (6.20) is an example of a *spectral method* solution. On the computer, $u(x, t)$ is approximated by

$$u_{N_F}(x,t) = \sum_{n=1}^{N_F} a_n(t) \sin(nx) \tag{6.21}$$

where N_F is the number of Fourier modes, and the error

$$\|u(x,t) - u_{N_F}(x,t)\| \sim e^{-N_F^2 t} \to 0 \tag{6.22}$$

as $N_F \to \infty$ faster than any finite power of $1/N_F$. This convergence is sometimes called infinite-order accuracy by advocates of spectral methods, but in practice, spectral methods simply achieve some high but finite degree of accuracy.

For initial conditions $u_0(x) = x(\pi - x)$ (cf. Fig. 6.1), (6.17) gives

$$a_n(0) = \frac{8}{n^3 \pi}, \quad n = 1, 3, 5, \ldots \tag{6.23}$$

for n odd and $a_n(0) = 0$ for n even.

6.3 Conservative Form

Many of the time-dependent PDEs in mathematical physics—and science in general—can be expressed as conservation laws, including the diffusion equation, the wave equations, Maxwell's equations, Burgers' equation, compressible fluid dynamics, the drift-diffusion model, the classical hydrodynamic model, general relativity, etc. *Conservative* numerical methods should *always* be used for discretizing conservation laws.

Linear and nonlinear diffusion in 3D can be written in the form of a microscopic conservation law for the conserved variable $u(\mathbf{x}, t)$:

$$\frac{\partial u}{\partial t} - \nabla \cdot (D(u) \nabla u) = \frac{\partial u}{\partial t} + \nabla \cdot \mathbf{f} = 0 \tag{6.24}$$

where $\mathbf{f} = \mathbf{f}(u, \nabla u) = -\nabla \cdot (D(u) \nabla u)$ is the flux of the conserved variable $u(\mathbf{x}, t)$.

To interpret the 3D microscopic conservation law (6.24), we derive the macroscopic conservation law for the conserved quantity $\overline{Q} = \int_V u \, dV$ by integrating over a volume V and using Gauss' Divergence Theorem:

$$\frac{d\overline{Q}}{dt} = \int_V \frac{\partial u}{\partial t} dV = -\int_{\partial V} \mathbf{f} \cdot \mathbf{da} = \text{inflow} - \text{outflow}. \tag{6.25}$$

For diffusion, $u(\mathbf{x}, t)$ is the number density of diffusing particles, $\mathbf{f}$ is the diffusing particle flux, and $\overline{Q}$ is the total number of diffusing particles in the volume V.

To discretize a 1D conservation law, we use the following:

Theorem *Theorem on Conservative Methods (1D Semi-Discrete): A numerical method for the 1D conservation law*

$$u_t + f(u, u_x)_x = 0 \tag{6.26}$$

is conservative *if it can be written in the semi-discrete form*

$$\frac{du_i(t)}{dt} + \frac{1}{\Delta x}\left(F_{i+\frac{1}{2}} - F_{i-\frac{1}{2}}\right) = 0 \tag{6.27}$$

where $F_{i+\frac{1}{2}}$ is the numerical flux function (defined by the numerical method) at $x_{i+\frac{1}{2}}$.

Proof Define $Q = \sum_{i=0}^{N} u_i \Delta x$. Then

$$\frac{dQ}{dt} = \sum_{i=0}^{N} \frac{du_i}{dt} \Delta x = -\sum_{i=0}^{N}\left(F_{i+\frac{1}{2}} - F_{i-\frac{1}{2}}\right) = F_{-\frac{1}{2}} - F_{N+\frac{1}{2}} = \text{inflow} - \text{outflow}. \tag{6.28}$$

Alternatively:

Theorem *Theorem on Conservative Methods (1D Fully Discrete): A one-step method for the 1D conservation law is conservative if it can be written in the fully discrete form*

$$u_i^{n+1} = u_i^n - \frac{\Delta t}{\Delta x}\left(F_{i+\frac{1}{2}} - F_{i-\frac{1}{2}}\right) \qquad (6.29)$$

where F may involve a combination of u^n and u^{n+1}.

Proof Define $Q^n = \sum_{i=0}^{N} u_i^n \Delta x$. Then

$$\frac{Q^{n+1} - Q^n}{\Delta t} = \sum_{i=0}^{N} \frac{u_i^{n+1} - u_i^n}{\Delta t} \Delta x = -\sum_{i=0}^{N}\left(F_{i+\frac{1}{2}} - F_{i-\frac{1}{2}}\right)$$

$$= F_{-\frac{1}{2}} - F_{N+\frac{1}{2}} = \text{inflow} - \text{outflow}. \qquad (6.30)$$

These ideas generalize to 3D in a straightforward way:

Theorem *Theorem on Conservative Methods (3D Semi-Discrete): A numerical method for the 3D conservation law*

$$u_t + \nabla \cdot \mathbf{f} = 0, \quad \mathbf{f} = (f, g, h) \qquad (6.31)$$

is conservative if it can be written (on a rectangular grid[1]) in the semi-discrete form

$$\frac{du_{ijk}}{dt} + \frac{F_{i+\frac{1}{2},j,k} - F_{i-\frac{1}{2},j,k}}{\Delta x} + \frac{G_{i,j+\frac{1}{2},k} - G_{i,j-\frac{1}{2},k}}{\Delta y} + \frac{H_{i,j,k+\frac{1}{2}} - H_{i,j,k-\frac{1}{2}}}{\Delta z} = 0$$

$$(6.32)$$

where $F_{i+\frac{1}{2},j,k}$ is the numerical flux for f at $\left(x_{i+\frac{1}{2}}, y_j, z_k\right)$, $G_{i,j+\frac{1}{2},k}$ is the numerical flux for g at $\left(x_i, y_{j+\frac{1}{2}}, z_k\right)$, and $H_{i,j,k+\frac{1}{2}}$ is the numerical flux for h at $\left(x_i, y_j, z_{k+\frac{1}{2}}\right)$.

A similar formulation applies to the 3D fully discrete form.

All of our numerical methods for the diffusion equation will be conservative, as well as all of our numerical methods for hyperbolic PDEs with the exception of the original first-order upwind method.[2]

---

[1] For more general grids, use the multidimensional finite volume approach in Sect. 6.12.
[2] The original upwind method is conservative for $u_t + (cu)_x = 0$ only if c does not change sign.

6.4 Parabolic Conservation Laws

For *parabolic conservation laws* like the nonlinear diffusion equation $u_t = (D(u)u_x)_x$ with $f = f(u, u_x) = -D(u)u_x$, the forward Euler, backward Euler, TR, and TRBDF2 methods are conservative with numerical flux

$$F_{i+\frac{1}{2}} = f_{i+\frac{1}{2}} = -D_{i+\frac{1}{2}}(u_x)_{i+\frac{1}{2}} = -\frac{1}{2}(D(u_i) + D(u_{i+1}))\frac{u_{i+1} - u_i}{\Delta x} \qquad (6.33)$$

taking appropriate linear combinations of F^n, F^{n+1}, and $F^{n+\gamma}$ for each numerical method. We can define $D_{i+\frac{1}{2}} = \frac{1}{2}(D(u_i) + D(u_{i+1}))$ or $D_{i+\frac{1}{2}} = D\left(\frac{1}{2}(u_i + u_{i+1})\right)$; the two definitions agree if $D(u) = au + b$. Thus, the conservative spatial discretization for the nonlinear diffusion equation is

$$\frac{du_i(t)}{dt} = \frac{1}{\Delta x^2}\left(D_{i+\frac{1}{2}}(u_{i+1} - u_i) - D_{i-\frac{1}{2}}(u_i - u_{i-1})\right). \qquad (6.34)$$

If $D = const$, we obtain the conservative spatial discretization for the linear diffusion equation $u_t = Du_{xx}$:

$$\frac{du_i(t)}{dt} = D\frac{u_{i+1} - 2u_i + u_{i-1}}{\Delta x^2}. \qquad (6.35)$$

Note Always discretize $(Du_x)_x$ (in 3D $\nabla \cdot (D\nabla u)$) directly instead of $Du_{xx} + D_x u_x$, the central discretization of which is nonconservative if $D_x \neq 0$ (see Problem 6.9).

6.5 Forward and Backward Euler Methods

The *backward Euler* method for the diffusion equation $u_t = Du_{xx}$

$$\frac{u_i^{n+1} - u_i^n}{\Delta t} = D\frac{u_{i+1}^{n+1} - 2u_i^{n+1} + u_{i-1}^{n+1}}{\Delta x^2} \qquad (6.36)$$

$$u_i^{n+1} = u_i^n + \frac{D\Delta t}{\Delta x^2}\left(u_{i+1}^{n+1} - 2u_i^{n+1} + u_{i-1}^{n+1}\right) \qquad (6.37)$$

(with $i = 1, 2, \ldots, N-1$ for Dirichlet boundary conditions[3]) is first-order accurate and A-stable and L-stable.

---

[3] Neumann boundary conditions are discussed in Sect. 6.10.2.

Equation (6.37) can be rewritten as

$$\left(I - D\Delta t\, \mathcal{D}^{(2)}\right) u^{n+1} = u^n \tag{6.38}$$

where I is the $(N-1) \times (N-1)$ identity matrix, $\mathcal{D}^{(2)} = \text{tridiag}[1, -2, 1]/h^2$ is the $(N-1) \times (N-1)$ second derivative matrix, and u^{n+1} and u^n are $N-1$ component vectors. Thus implicit methods (backward Euler, TR, TRBDF2) for the diffusion equation will involve a linear solve at each timestep.

The program **diffusion0.m** implements the backward Euler method (6.38) for the diffusion equation in one line using array operations ($D = 1$):

```
u = (speye(N-1) - dt*D2)\u; % backward Euler
```

Consistency: The LTE ($= \Delta t\, \tau$) is given by[4]

$$u(t + \Delta t)$$
$$= u + \frac{D\Delta t}{h^2}(u(x+h, t+\Delta t) - 2u(x, t+\Delta t) + u(x-h, t+\Delta t)) + \Delta t\, \tau. \tag{6.39}$$

Taylor expanding, we get

$$u + \Delta t\, u_t + \frac{\Delta t^2}{2} u_{tt} + \cdots = u + D\Delta t\left(u_{xx} + \Delta t\, u_{xxt} + \frac{h^2}{12} u_{xxxx} + \cdots\right) + \Delta t\, \tau \tag{6.40}$$

$$\tau = -\frac{\Delta t}{2} u_{tt} - \frac{Dh^2}{12} u_{xxxx} + \cdots \tag{6.41}$$

Thus backward Euler is first-order accurate, or more precisely, first-order accurate in time and second-order accurate in space.

Stability is treated in the next section.

The *forward Euler* method for the diffusion equation $u_t = Du_{xx}$

$$u_i^{n+1} = u_i^n + \frac{D\Delta t}{\Delta x^2}(u_{i+1}^n - 2u_i^n + u_{i-1}^n) \tag{6.42}$$

is first-order accurate and conditionally stable, i.e., stable only if $\Delta t \leq \Delta x^2/(2D)$ (see Problem 6.3).

---

[4] Some authors ([26], e.g.) call τ the local truncation error for PDEs. Our definition is consistent with the LTE (2.31) for ODE initial value problems. Of course everyone agrees that backward Euler is globally first-order accurate.

6.6 Implicit vs. Explicit Methods for the Diffusion Equation

Even though implicit methods for the diffusion equation (backward Euler, TR, and TRBDF2) require linear solves like (6.38) at each timestep, these methods can take $\Delta t \sim \Delta x$ instead of the explicit timestep limit $\Delta t \sim \Delta x^2$, so not only are the implicit methods more stable, but they also are of near optimal efficiency in 1D, requiring $\sim N$ flops per timestep for the tridiagonal solve and thus $\sim N \times N$ flops overall, while explicit methods require $\sim N \times N^2$ flops overall.

In 2D and 3D, PCG or GMRES (see Chap. 7) can be used for the linear solves in the implicit methods, requiring $N_{iter} \sim N^{d+1}$ flops in d dimensions, and thus $\sim N^{d+1} \times N$ flops overall. Explicit methods in 2D and 3D would have the same complexity ($\sim N^d \times N^2$), but due to stability concerns (especially for nonlinear parabolic PDEs), we always recommend using TRBDF2.

6.7 Von Neumann Stability Analysis

To analyze the stability of numerical methods for time-dependent PDEs, we will use von Neumann's Fourier analysis (see Sect. 2.7). Set the initial data u_j^n at time level n equal to a single Fourier mode $u_j^n = e^{ikx_j}$, where $k = 2\pi/\lambda$ is the wave number and λ is the wavelength of the Fourier mode, and then derive the growth factor $u_j^{n+1} = G(k) u_j^n$ for this mode. We require $|G(k)| \leq 1$ for all k.

For backward Euler, we find

$$0 < G(k) = \frac{1}{1 + \frac{4D\Delta t}{h^2} \sin^2\left(\frac{kh}{2}\right)} \leq 1 \qquad (6.43)$$

so backward Euler is A-stable. (Note that Δt is always a positive number: $0 < \Delta t < \infty$.) Backward Euler is L-stable as well, since $\lim_{\Delta t \to \infty} |G| = 0$ almost everywhere (there are isolated values of k where $G(k) = 1$ independently of Δt).

Derivation of $G(k)$ Set $u_j^n = e^{ikx_j}$ and $u_j^{n+1} = G(k) u_j^n = G(k) e^{ikx_j}$ in the backward Euler method:

$$u_j^{n+1} = G e^{ikx_j} = u_j^n + \frac{D\Delta t}{h^2}(u_{j+1}^{n+1} - 2u_j^{n+1} + u_{j-1}^{n+1})$$

$$= e^{ikx_j} + \frac{D\Delta t}{h^2} G \left(e^{ikx_{j+1}} - 2e^{ikx_j} + e^{ikx_{j-1}}\right)$$

$$= e^{ikx_j} + \frac{D\Delta t}{h^2} \left(e^{ikh} - 2 + e^{-ikh}\right) G e^{ikx_j}$$

$$= e^{ikx_j} + \frac{2D\Delta t}{h^2} (\cos(kh) - 1) G e^{ikx_j}$$

$$= e^{ikx_j} - \frac{4D\Delta t}{h^2}\sin^2\left(\frac{kh}{2}\right)Ge^{ikx_j}. \tag{6.44}$$

Now solving for G yields

$$G = 1 - \frac{4D\Delta t}{h^2}\sin^2\left(\frac{kh}{2}\right)G \tag{6.45}$$

$$G(k) = \frac{1}{1 + \frac{4D\Delta t}{h^2}\sin^2\left(\frac{kh}{2}\right)}. \tag{6.46}$$

Stability of forward Euler Repeating an analogous derivation, we obtain

$$-1 \le G(k) = 1 - \frac{4D\Delta t}{h^2}\sin^2\left(\frac{kh}{2}\right) \le 1 \tag{6.47}$$

which implies $\Delta t \le h^2/(2D)$.

6.8 Trapezoidal Rule Method

The trapezoidal rule method for the heat/diffusion equation is also known historically as the Crank-Nicolson method.

The TR method for the diffusion equation $u_t = Du_{xx}$

$$u_i^{n+1} = u_i^n + \frac{D\Delta t}{2\Delta x^2}(u_{i+1}^n - 2u_i^n + u_{i-1}^n + u_{i+1}^{n+1} - 2u_i^{n+1} + u_{i-1}^{n+1}) \tag{6.48}$$

is second-order accurate and A-stable but not L-stable for the diffusion equation (see Problems 6.4 and 6.5).

6.9 Relationship of Parabolic PDEs to ODE IVPs

Parabolic PDEs are closely related to stiff ODE initial value problems. In fact, if the diffusion equation $u_t = Du_{xx}$ is discretized in space (the semi-discrete problem) using three-point central differences, the errors τ and growth factors $G(k)$ for the original PDE are related to those for the ODE $du/dt = -\alpha u$, $\alpha > 0$, for forward Euler, backward Euler, TR, TRBDF2, etc., by the transcription:

$$\tau = \tau_{ODE} - \frac{Dh^2}{12}u_{xxxx} + \cdots \tag{6.49}$$

and

$$G(k) = G_{ODE}(\alpha \Delta t), \quad \alpha = \frac{4D}{h^2} \sin^2\left(\frac{kh}{2}\right). \tag{6.50}$$

6.10 Boundary Conditions for Parabolic/Elliptic PDEs

We discuss the 1D case with a three-point stencil for simplicity; the extensions to wider stencils and to multidimensional Dirichlet boundary conditions are straightforward. Implementation of Neumann boundary conditions for multidimensional parabolic/elliptic PDEs will be discussed in Sect. 6.12 on the finite volume method.

Take $N + 1$ grid points on a spatial interval $[a, b]$, with $N \Delta x = b - a$. Label the solution values $u_0, u_1, \ldots, u_N$. The discussion of boundary conditions applies equally well to a scalar PDE where each u_i is a scalar or to a PDE system where each u_i is a vector.

Mathematically appropriate boundary conditions for parabolic/elliptic PDEs are Dirichlet (already used in our linear diffusion programs) or Neumann (implemented in our nonlinear diffusion program in Sect. 6.11) or Robin (a linear combination of Dirichlet and Neumann conditions).

6.10.1 Dirichlet Boundary Conditions

Simply set $u_0 = u_a$ and $u_N = u_b$, and implement the PDE on the interior grid points $i = 1, 2, \ldots, N - 1$. Note that u_a and u_b can be functions of t.

6.10.2 Neumann Boundary Conditions

The best way to implement Neumann boundary conditions $u_x(a, t) \equiv u_{xa}$ and $u_x(b, t) \equiv u_{xb}$ is by allotting *ghost points* to the left and right of the boundaries. For a three-point stencil, add ghost points u_{-1} and u_{N+1} (see Fig. 6.2) with

$$u_x(a, t) \approx \frac{u_1 - u_{-1}}{2\Delta x} = u_{xa}, \quad u_{-1} = u_1 - 2u_{xa}\Delta x \tag{6.51}$$

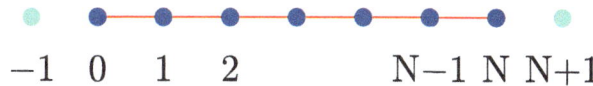

Fig. 6.2 Neumann boundary conditions

$$u_x(b,t) \approx \frac{u_{N+1} - u_{N-1}}{2\Delta x} = u_{xb}, \quad u_{N+1} = u_{N-1} + 2u_{xb}\Delta x \tag{6.52}$$

and then implement the PDE at the grid points $i = 0, 1, \ldots, N$. With homogeneous Neumann boundary conditions, the ghost points simply replicate the first and last interior point values $u_{-1} = u_1$ and $u_{N+1} = u_{N-1}$.

Note that u_{xa} and u_{xb} can be functions of t and that the Neumann problem with $N + 1$ unknowns is larger by 2 in 1D than the Dirichlet problem with $N - 1$ unknowns. Our MATLAB program **diffusionNBC.m** implements the TRBDF2 method for the linear diffusion equation with dynamic timestep and homogeneous Neumann boundary conditions. The reader can check that with homogeneous Neumann boundary conditions, the total number of particles (or total amount of heat energy) remains constant in the domain (see Problem 6.12).

6.11 Nonlinear Diffusion: TRBDF2 and Newton's Method

TRBDF2 is the method of choice for linear and nonlinear parabolic PDEs, since it is second-order accurate and L-stable. For linear diffusion, skip the Newton iteration steps below. Computed solutions using our MATLAB TRBDF2 codes **diffusion2.m** for the 1D linear diffusion equation, **nonlin_diffusion.m** for the 1D nonlinear diffusion equation, and **diffusion2D.m** for the 2D linear diffusion equation are presented in Figs. 6.1, 6.4, 6.5, and 6.6.

6.11.1 TRBDF2 Revisited

We can make an exact transcription of Sect. 3.9 on TRBDF2 for ODE initial value problems by spatially discretizing the nonlinear diffusion equation [23]

$$\frac{\partial u}{\partial t} = \nabla \cdot (D(u)\nabla u) \tag{6.53}$$

using the conservative numerical flux given in Sect. 6.4

$$F_{i+\frac{1}{2}} = -D_{i+\frac{1}{2}}(u_x)_{i+\frac{1}{2}} = -\frac{1}{2}(D(u_i) + D(u_{i+1}))\frac{u_{i+1} - u_i}{\Delta x} \tag{6.54}$$

where

$$D_{i\pm\frac{1}{2}} = \frac{1}{2}(D_i + D_{i\pm 1}). \tag{6.55}$$

6.11 Nonlinear Diffusion: TRBDF2 and Newton's Method

Thus, the conservative spatial discretization for the nonlinear diffusion equation generates the semi-discrete problem[5]

$$\frac{du_i(t)}{dt} = \frac{1}{\Delta x^2}\left(D_{i+\frac{1}{2}}(u_{i+1}-u_i) - D_{i-\frac{1}{2}}(u_i - u_{i-1})\right) \equiv f_i(u). \tag{6.56}$$

Then to integrate[6] $du/dt = f(u)$ from $t = t_n$ to $t_{n+1} = t_n + \Delta t$, first apply the trapezoidal rule (TR) to advance the solution from t_n to $t_{n+\gamma} = t_n + \gamma \Delta t$:

$$u_{n+\gamma} - \gamma \frac{\Delta t}{2} f_{n+\gamma} = u_n + \gamma \frac{\Delta t}{2} f_n \tag{6.57}$$

and then use the second-order backward differentiation formula (BDF2) to advance the solution to t_{n+1}:

$$u_{n+1} - \frac{1-\gamma}{2-\gamma}\Delta t\, f_{n+1} = \frac{1}{\gamma(2-\gamma)}u_{n+\gamma} - \frac{(1-\gamma)^2}{\gamma(2-\gamma)}u_n. \tag{6.58}$$

Again, we linearize f_{n+1} in (6.58) (and similarly $f_{n+\gamma}$ in (6.57)) iteratively by setting the new iterative solution to

$$u_{n+1}^{(k+1)} = u_{n+1}^{(k)} + \Delta u^{(k)}, \quad u_{n+1}^{(0)} = u_{n+\gamma} \tag{6.59}$$

and approximating

$$f_{n+1}^{(k+1)} = f_{n+1}^{(k)} + \left(\frac{\partial f}{\partial u}\right)_{n+1}^{(k)} \Delta u^{(k)} \tag{6.60}$$

where $k = 0, 1, 2, \ldots$ labels the Newton iterations. At each TR or BDF2 partial step, we iterate until the Newton method converges.

The Newton equation for the TR partial step is

$$\left[I - \gamma \frac{\Delta t}{2}\left(\frac{\partial f}{\partial u}\right)_{n+\gamma}^{(k)}\right]\Delta u^{(k)} = -\left(u_{n+\gamma}^{(k)} - u_n\right) + \gamma \frac{\Delta t}{2}\left(f_{n+\gamma}^{(k)} + f_n\right) \equiv -R_{\text{TR}} \tag{6.61}$$

where R_{TR} is the residual for (6.57).

---

[5] In this section, $f(u)$ is the right-hand side of $du/dt = f(u)$ and *not* the flux function.
[6] We summarize the TRBDF2 formulas from Sect. 3.9 for the reader's convenience.

The Newton equation for the BDF2 partial step is

$$\left[I - \frac{1-\gamma}{2-\gamma}\Delta t \left(\frac{\partial f}{\partial u}\right)^{(k)}_{n+1}\right]\Delta u^{(k)}$$

$$= -\left(u^{(k)}_{n+1} - \frac{1}{\gamma(2-\gamma)}u_{n+\gamma} + \frac{(1-\gamma)^2}{\gamma(2-\gamma)}u_n\right) + \frac{1-\gamma}{2-\gamma}\Delta t f^{(k)}_{n+1} \equiv -R_{\text{BDF2}} \tag{6.62}$$

where R_{BDF2} is the residual for (6.58).

The divided-difference formula for the timestep selection is implemented in Sect. 6.11.2.

6.11.2 Dynamic Timestep

The timestep size Δt is adjusted dynamically within a window $[\Delta t_{\min}, \Delta t_{\max}]$ by monitoring a divided-difference estimate of the local error e_l:

$$e_l = k_\gamma \Delta t^3 u''' \approx 2k_\gamma \Delta t \left(\frac{1}{\gamma}f_n - \frac{1}{\gamma(1-\gamma)}f_{n+\gamma} + \frac{1}{1-\gamma}f_{n+1}\right) \tag{6.63}$$

where

$$k_\gamma = \frac{-3\gamma^2 + 4\gamma - 2}{12(2-\gamma)}. \tag{6.64}$$

We recommend the value $\gamma = 2 - \sqrt{2} \approx 0.59$, which minimizes the norm of the local error, but the MATLAB value $\gamma = \frac{1}{2}$ is also fine.

6.11.3 Accuracy and Stability

Using the formalism of Sect. 6.9, the global truncation error for TRBDF2 for the linear diffusion equation $u_t = D u_{xx}$ is given by

$$\tau = k_\gamma \Delta t^2 u_{ttt} - \frac{Dh^2}{12}u_{xxxx} + \cdots \tag{6.65}$$

and the growth factor by

6.11 Nonlinear Diffusion: TRBDF2 and Newton's Method

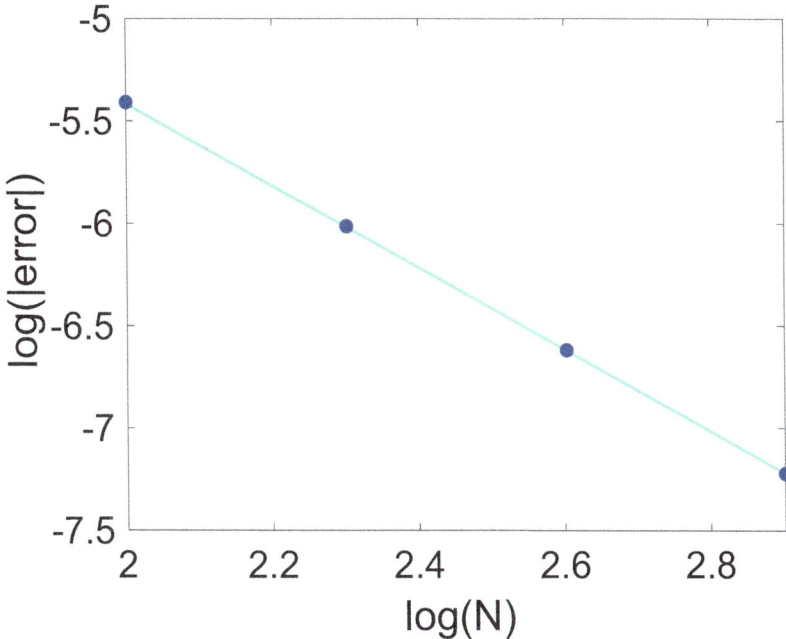

Fig. 6.3 Numerical errors $\log_{10}(\|u_{exact} - u_{comp}\|_1)$ vs. $\log_{10}(N)$ for the linear diffusion equation comparing the TRBDF2 solution from **diffusion1.m** with the "exact" Fourier solution for $N = 100, 200, 400,$ and $800\Delta x$ at $t = 0.5$. The *straight line* is the theoretical value with log-log slope of -2

$$G(\Delta t, \gamma) = \frac{\frac{1 - \frac{\gamma \alpha \Delta t}{2}}{1 + \frac{\gamma \alpha \Delta t}{2}} - (1-\gamma)^2}{\gamma(2-\gamma) + \gamma(1-\gamma)\alpha \Delta t}, \quad \alpha = \frac{4D}{h^2}\sin^2\left(\frac{kh}{2}\right) \quad (6.66)$$

(see Problem 6.6). L-stability follows by Problem 3.18 since $|G| \leq 1$ and $\lim_{\Delta t \to \infty} G = 0$.

Figure 6.3 demonstrates that our implementation of TRBDF2 for the linear diffusion equation with homogeneous Dirichlet boundary conditions and initial conditions $u_0(x) = 1 - x^2$ on $-1 \leq x \leq 1$ is second-order accurate by comparing the computed solution with the "exact" Fourier solution on various spatial grids.

Problem 6.13 shows that our implementation of TRBDF2 and Newton's method in **nonlin_diffusion.m** is second-order accurate.

6.11.4 Jacobian $\partial f / \partial u$

For the TR and BDF2 partial steps, we need the Jacobian $\partial f_i / \partial u_j$, with f_i given in (6.56).

Since the diffusion coefficient $D_i = D(u_i)$ is a local (pointwise) function of the concentration,

$$\frac{\partial D_{i \pm \frac{1}{2}}}{\partial u_j} = \frac{1}{2} \left(D'_i \delta_{ij} + D'_{i \pm 1} \delta_{i \pm 1, j} \right), \quad D'_i = \frac{dD}{du}(u_i). \tag{6.67}$$

Thus, the explicit form of the Jacobian is

$$\Delta x^2 \frac{\partial f_i}{\partial u_j} = \delta_{ij} \left(\frac{1}{2} D'_i (u_{i+1} + u_{i-1} - 2u_i) - D_{i+1} - D_{i-1} \right)$$

$$+ \delta_{i+1,j} \left(\frac{1}{2} D'_{i+1} (u_{i+1} - u_i) + D_{i+\frac{1}{2}} \right)$$

$$+ \delta_{i-1,j} \left(\frac{1}{2} D'_{i-1} (u_i - u_{i-1}) + D_{i-\frac{1}{2}} \right). \tag{6.68}$$

The Jacobian $\partial f_i / \partial u_j$ can also be computed by finite differencing f_i with respect to Δu_j using the method at the end of Sect. 3.8, but that is not necessary here, since the Jacobian is so simple. The explicit analytical Jacobian is provided in **nonlin_diffusion.m**.

As a realistic example in **nonlin_diffusion.m**, we will take $D(u) = au + b$ (a and b positive), with $D'(u) = a$.

6.11.5 TRBDF2 Simulations of Nonlinear Diffusion

TRBDF2 computations using **nonlin_diffusion.m** for the nonlinear and *linear* diffusion equations with homogeneous Neumann boundary conditions and a Gaussian implant are shown in Figs. 6.4 and 6.5. Note how the nonlinear solution is steeper and less smoothed out than the linear solution. It is exactly this kind of steeper and less smoothed-out diffusion that is needed to make reliable devices on a semiconductor wafer.

With homogeneous Neumann boundary conditions, the total amount $Q(t)$ of dopant in the domain should remain constant in time in exact arithmetic, and in floating point, arithmetic should remain constant up to roundoff error. By running

```
nonlin_diffusion(100,0.5)
```

the reader can verify that

$$Q(t) = \int u(x,t) \, dx \approx Q^n = \frac{1}{2} \sum_{i=1}^{N} (u_i^n + u_{i+1}^n) \Delta x \tag{6.69}$$

6.12 Finite Volume Method for Multidimensional PDEs

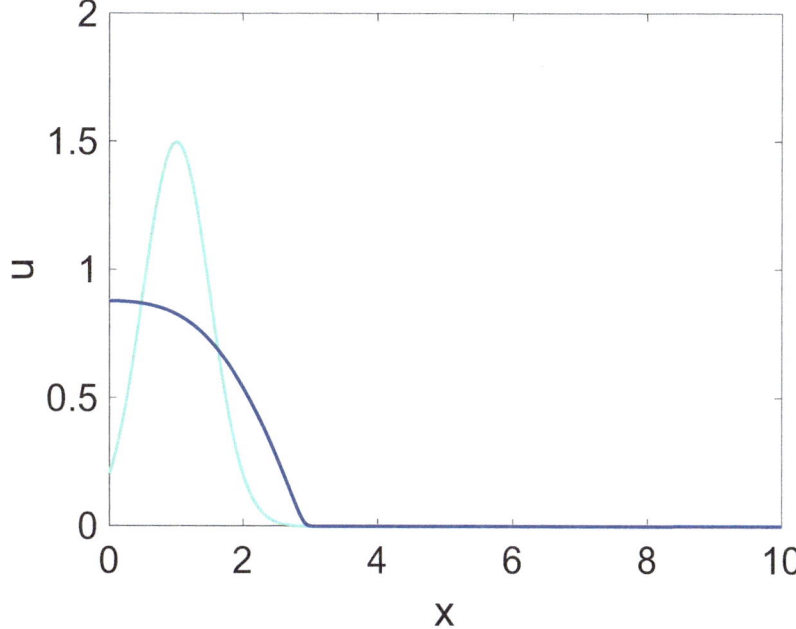

Fig. 6.4 Gaussian initial conditions (*peaked curve*) and computed solution at $t = 0.4$ using **nonlin_diffusion.m** (TRBDF2) for the nonlinear diffusion equation ($D = 1.5u + 0.02$) on a 1 micron domain with $200\Delta x$ and homogeneous Neumann boundary conditions

(MATLAB labeling) remains constant in time up to roundoff error.

The **nonlin_diffusion.m** code is quite compact with array operations. The reader is invited to imagine how complex the code would become if for-loops are used instead of array operations.

6.12 Finite Volume Method for Multidimensional PDEs

In multiple spatial dimensions, Gauss' Divergence Theorem can be used to discretize spatial operators like $\nabla \cdot \mathbf{f}$ or $-\nabla \cdot (D(u)\nabla u)$. This finite volume approach (or *box method*) for spatial operators works well for parabolic and elliptic PDEs but generally spreads out waves too much for hyperbolic PDEs (see Sect. 8.22). However, the *space-time* finite volume method is the basis for many conservative discretizations of hyperbolic conservation laws (see Sect. 8.21).

For simplicity, we describe the finite volume discretization first on a rectangular grid with grid spacings $\Delta x, \Delta y, \Delta z$. Applying Gauss' Divergence Theorem to a grid volume $V = \Delta x \Delta y \Delta z$, the conservation law $u_t + \nabla \cdot \mathbf{f} = 0$, where $\mathbf{f} = (f, g, h)$, becomes

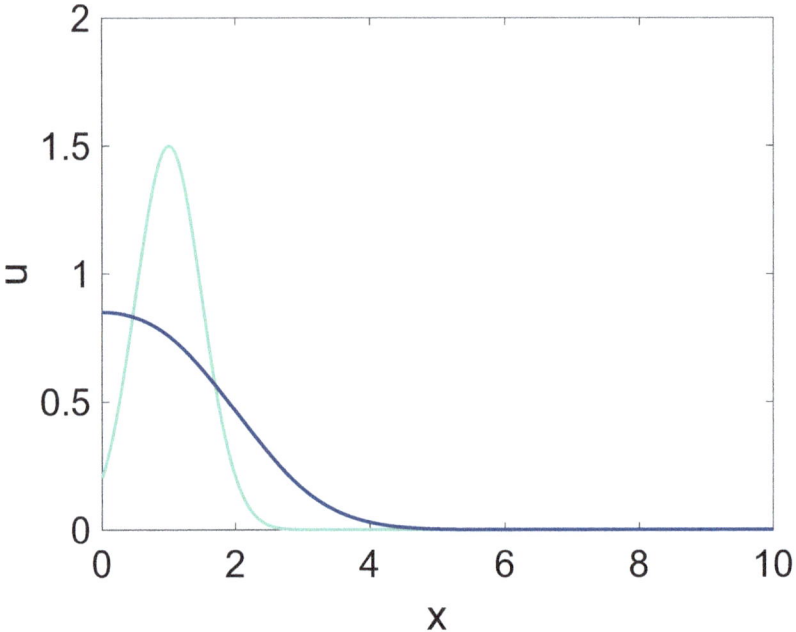

Fig. 6.5 Gaussian initial conditions (*peaked curve*) and computed solution at $t = 0.4$ using **nonlin_diffusion.m** (TRBDF2) for the *linear* diffusion equation ($D = 1.5$) on a 1 micron domain with $200\Delta x$ and homogeneous Neumann boundary conditions

$$\frac{du_{ijk}}{dt} \equiv \frac{1}{V}\int dV\,(u_t)_{ij} = -\frac{1}{V}\int da\,\hat{\mathbf{n}}\cdot\mathbf{f}$$

$$= -\frac{f_{i+\frac{1}{2},j,k} - f_{i-\frac{1}{2},j,k}}{\Delta x} - \frac{g_{i,j+\frac{1}{2},k} - g_{i,j-\frac{1}{2},k}}{\Delta y} - \frac{h_{i,j,k+\frac{1}{2}} - h_{i,j,k-\frac{1}{2}}}{\Delta z} \quad (6.70)$$

where $\hat{\mathbf{n}}$ is a unit outward-pointing normal vector to the boundary of V. The terms on the right-hand side like $f_{i+\frac{1}{2},j,k}$ are approximated by

$$f_{i+\frac{1}{2},j,k} \approx \frac{1}{2}(f_{i+1,j,k} + f_{ijk}) \quad (6.71)$$

etc. If $\mathbf{f} = -D(u)\nabla u$, then

$$f_{i+\frac{1}{2},j,k} \approx -\frac{1}{2}\left(D(u_{i+1,j,k}) + D(u_{ijk})\right)\frac{u_{i+1,j,k} - u_{ijk}}{\Delta x} \quad (6.72)$$

etc. These discretizations then agree with the second-order accurate finite difference formulas—see **diffusion2D.m** and the 2D simulations in Fig. 6.6.

6.12 Finite Volume Method for Multidimensional PDEs

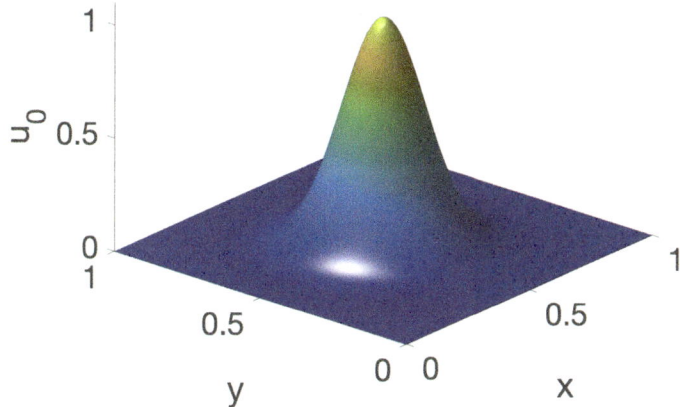

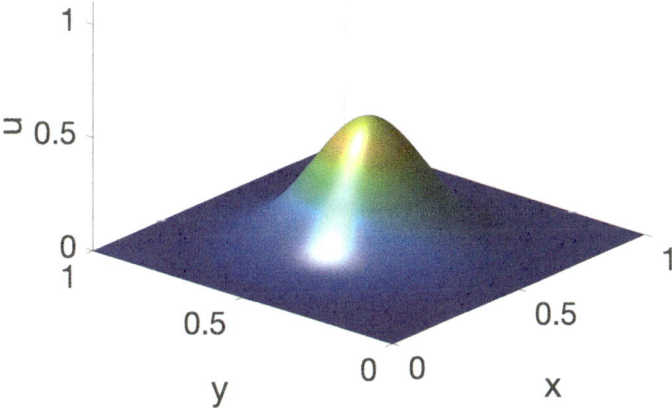

Fig. 6.6 Gaussian initial conditions and TRBDF2 computed solution at $t = 0.005$ for linear diffusion in 2D using **diffusion2D.m** on a $100\Delta x \times 100\Delta y$ grid

The finite volume method generalizes to 2D and 3D non-rectangular grids and yields a method for discretizing Neumann boundary conditions along slanted boundaries that do not align with the grid axes. For more general boxes, $\mathbf{f} \cdot \hat{\mathbf{n}}$ and $\hat{\mathbf{n}} \cdot \nabla u$ can be evaluated from whatever grid point values are available along each portion of the boundary of the box.

It is straightforward to apply the finite volume discretization to general 2D triangularizations, but the extension to 3D tetrahedra requires more sophisticated geometrical machinery, and software packages are recommended.

Problems

6.1 Show the eigenvalues of the $n \times n$ matrix second derivative matrix $\mathcal{D}^{(2)}$ are

$$\lambda_j = \frac{2}{h^2}(\cos(j\pi h) - 1), \quad j = 1, 2, \ldots, n$$

with $h = 1/(n+1)$. *Hint:* $\mathcal{D}^{(2)} = \text{tridiag}[1, -2, 1]/h^2$. Show

$$\left(\mathcal{D}^{(2)} v^{(j)}\right)_k = \lambda_j v_k^{(j)}, \quad j, k = 1, 2, \ldots, n$$

where the eigenvector

$$v_k^{(j)} = \sin(j\pi k h).$$

6.2 Show the general solution to the diffusion equation

$$u(x, t) = \int_{-\infty}^{\infty} K(x - y, t) u_0(y) \, dy$$

where the fundamental solution or kernel

$$K(x, t) = \frac{1}{\sqrt{4\pi Dt}} \exp\left\{-\frac{x^2}{4Dt}\right\}$$

does in fact satisfy $u_t = D u_{xx}$ with $u(x, t = 0) = u_0(x)$. (The Leibniz integral rule allows us to interchange the order of differentiation and integration.) *Hint:* First show K satisfies the diffusion equation. Then argue that u satisfies the diffusion equation. Finally show u satisfies the initial conditions $u(x, t = 0) = u_0(x)$.

6.3 Prove the forward Euler method (6.42) for the diffusion equation $u_t = D u_{xx}$ is (a) first-order accurate (using the definition of the LTE = $\Delta t\, \tau$) and (b) conditionally stable. What is the restriction on Δt? *Hints:* For stability, set $u_j^n = e^{ikx_j}$ and derive $u_j^{n+1} = G(k) u_j^n$. Then require $|G(k)| \leq 1$ for all wave numbers k.

6.4 Show the TR method (6.48) for the diffusion equation $u_t = D u_{xx}$ is second-order accurate, using the definition of the local truncation error: i.e., show the global error $\tau = c_1 \Delta t^2 + c_2 \Delta x^2 + \cdots$, where "$\cdots$" means higher-order terms in Δt and Δx. (Fill in the values of c_1 and c_2.) *Hint:* Use the facts that

Problems

$$\frac{u(x+h) - 2u(x) + u(x-h)}{h^2} = u_{xx} + \frac{h^2}{12}u_{xxxx} + \cdots$$

$$u_{xx}(t + \Delta t) = u_{xx} + \Delta t \, u_{xxt} + \frac{\Delta t^2}{2} u_{xxtt} + \cdots$$

and that by virtue of the PDE $u_t = Du_{xx}$, $u_{xxt} = u_{tt}/D$ and $u_{xxtt} = u_{ttt}/D$.

6.5 Show the growth factor for the TR method for the diffusion equation $u_t = Du_{xx}$ is

$$G(k) = \frac{1 - \frac{2D\Delta t}{h^2}\sin^2\left(\frac{kh}{2}\right)}{1 + \frac{2D\Delta t}{h^2}\sin^2\left(\frac{kh}{2}\right)}.$$

Then show TR is A-stable but not L-stable. *Hints:* For stability, set $u_j^n = e^{ikx_j}$ and derive $u_j^{n+1} = G(k)u_j^n$. Then for A-stability require $|G(k)| \leq 1$ for all wave numbers k and any $\Delta t > 0$. For L-stability, *also* require $\lim_{\Delta t \to \infty} G(k) = 0$.

6.6 Review Problems 3.17–3.21 for TRBDF2 for the semi-discrete ODE (method of lines) problem $du/dt = f(u)$. Show directly for $u_t = Du_{xx}$, without using the transcriptions in Sect. 6.9, that

$$\tau = k_\gamma \Delta t^2 u_{ttt} - \frac{Dh^2}{12}u_{xxxx} + \cdots$$

and that $G(k)$ has the same form as for the ODE $du/dt = -\alpha u$, $\alpha > 0$, with

$$\alpha = \frac{4D}{h^2}\sin^2\left(\frac{kh}{2}\right).$$

6.7 Derive the Fourier solution to the initial/boundary value problem for the diffusion equation with homogeneous Neumann boundary conditions

$$u_t = u_{xx}, \quad u_x(0,t) = 0 = u_x(\pi,t) \text{ for } t > 0, \quad u(x,t=0) = u_0(x)$$

by making a Fourier cosine expansion (since it automatically satisfies the boundary conditions)

$$u(x,t) = \sum_{n=0}^{\infty} a_n(t) \cos(nx)$$

and solving for $a_n(t)$.

6.8 The nonlinear diffusion equation is

$$\frac{\partial u}{\partial t} = \frac{\partial}{\partial x}\left(D(u)\frac{\partial u}{\partial x}\right).$$

Discretize this PDE using backward Euler and central differences directly (yielding a *conservative method*); i.e., *do not* use the product rule for the right-hand side, but discretize first the outer ∂_x and then the inner ∂_x as

$$\left(\frac{\partial f}{\partial x}\right)_i = \frac{f_{i+\frac{1}{2}} - f_{i-\frac{1}{2}}}{h}$$

and

$$\left(\frac{\partial g}{\partial x}\right)_{i+\frac{1}{2}} = \frac{g_{i+i} - g_i}{h}, \quad \left(\frac{\partial g}{\partial x}\right)_{i-\frac{1}{2}} = \frac{g_i - g_{i-1}}{h}$$

where f and g are any quantities that are differentiated. You can use u and D at $i-1, i-\frac{1}{2}, i, i+\frac{1}{2}, i+1$.

6.9 Show the second-order accurate central discretization

$$u_i^{n+1} = u_i^n + \frac{\Delta t}{\Delta x^2}\left(\frac{1}{4}(D_{i+1} - D_{i-1})(u_{i+1} - u_{i-1}) + D_i(u_{i+1} - 2u_i + u_{i-1})\right)$$

of $u_t = (D(u)u_x)_x$ written out as $D_x u_x + D u_{xx}$ is nonconservative. (You can use $Q^n = \sum_{i=0}^N u_i^n \Delta x$ and $N = 4$. Then show the terms multiplying u_2 in $(Q^{n+1} - Q^n)/\Delta t$ do not cancel unless D is constant.)

6.10 (a) Using the trapezoidal rule integration method to approximate

$$\bar{Q}(t) = \int u(x,t)\,dx \approx Q^n = \sum_{i=0}^{N-1} \frac{1}{2}(u_i^n + u_{i+1}^n)\Delta x$$

show discretizations of $u_t + f_x = 0$ of the form

$$u_i^{n+1} = u_i^n - \frac{\Delta t}{\Delta x}\left(F_{i+\frac{1}{2}} - F_{i-\frac{1}{2}}\right)$$

are conservative.

(b) For nonlinear diffusion, $f(u) = -D(u)u_x$ and $F_{i+\frac{1}{2}} = -D_{i+\frac{1}{2}}(u_{i+1} - u_i)/\Delta x$. Show if homogeneous Neumann boundary conditions $u_x(a,t) = 0 = u_x(b,t)$ are imposed; Q is constant in time.

6.11 Show **diffusion2.m** (TRBDF2) converges under mesh refinement by comparing and contrasting simulations with $N = 50, 100,$ and $200\Delta x$ at $t = 0.25$.

Problems

6.12 Add code to **diffusionNBC.m** to calculate the total number of particles in the domain

$$Q(t) = \int_{-1}^{1} u(x,t)\,dx \approx \frac{1}{2} \sum_{i=1}^{N} (u_i^n + u_{i+1}^n)\,\Delta x$$

(MATLAB labeling), and show $Q(t)$ remains constant in time up to roundoff error. Check Q at $t = 0, 0.5, 1$, and 2. Discuss the behavior of the solution in time.

6.13 For **nonlin_diffusion.m**, an exact solution is not available. Estimate the order of accuracy for the TRBDF2 method for nonlinear diffusion by using a fine grid solution u_{fine} to stand in for the exact solution. Verify that our implementation achieves second-order accuracy in Δx and Δt by comparing computed solutions at $t = 0.4$ on spatial grids with $N = 100, 200$, and $400\Delta x$ with a fine grid solution with $800\Delta x$. Take a fixed timestep with $\Delta t = N\Delta x^2 / \max\{D(u_0)\}$, where the maximum is over all grid points at $t = 0$.

6.14 Simulate nonlinear diffusion using **nonlin_diffusion.m** with $100\Delta x$. Plot (in one figure) $u(x,t)$ for $t = 0, 500, 1000, 1500$, and 2000 s. Discuss your results. *Hint:* $t_{comp} = 0, 0.5, 1, 1.5, 2$.

6.15 *Project:* Install dynamic timestepping in **diffusion2D.m** for 2D linear diffusion equation.

Chapter 7
Numerical Methods for Elliptic PDEs

Poisson's equation governs electrostatics with free charges and Newtonian gravitation; Laplace's equation governs electrostatics when there are no free charges and Newtonian gravitation in empty space, as well as many other applications in science and engineering like potential flow in vortex-free fluid dynamics. Laplace's and Poisson's equations are our primary examples of *elliptic* PDEs.

The 3D Poisson equation (for electrostatics) is

$$\nabla \cdot (\varepsilon \nabla u) = -\rho(\mathbf{x}) \tag{7.1}$$

where $\rho(\mathbf{x})$ is the free charge density and ε is the permittivity (dielectric coefficient), which may in general depend on $\mathbf{x}$ and/or u. For Newtonian gravity, the 3D gravitational Poisson equation is

$$\nabla^2 u = 4\pi G \rho(\mathbf{x}) \tag{7.2}$$

where $\rho(\mathbf{x})$ is the mass density and G is Newton's gravitational constant. In 3D, Laplace's equation (for electrostatics) is

$$\nabla \cdot (\varepsilon \nabla u) = 0. \tag{7.3}$$

Just as for parabolic PDEs (see Sects. 5.1 and 6.10), mathematically appropriate boundary conditions for elliptic PDEs are Dirichlet ($u(\mathbf{x}_\mathcal{B}, t)$ is specified on the boundary $\mathbf{x}_\mathcal{B}$) or Neumann ($\hat{\mathbf{n}} \cdot \nabla u(\mathbf{x}_\mathcal{B}, t)$ is specified on the boundary) or Robin (a linear combination of Dirichlet and Neumann conditions is specified on the boundary). Again, Dirichlet or Neumann or Robin boundary conditions can be specified on different parts of the boundary.

In 2D (for simplicity) with a constant ε, Laplace's equation is

$$\nabla^2 u = \frac{\partial^2 u}{\partial x^2} + \frac{\partial^2 u}{\partial y^2} = 0 \tag{7.4}$$

where $u(x, y)$ is the electrostatic potential (measured in volts) with the electric field $\mathbf{E} = -\nabla u$, or the velocity potential with $\mathbf{v} = \nabla u$, etc. The partial derivatives can be abbreviated by subscripts: $u_{xx} + u_{yy} = 0$. Discretizing the second derivatives, we can convert Laplace's equation into a matrix-vector equation $Au = b$, where u is the solution vector and b incorporates the boundary conditions (see Fig. 7.1). Note the similarity with discretizing linear ODE boundary value problems.

The equation $Au = b$ can by solved directly in 1D or 2D by Gaussian elimination, but memory usage in 3D requires an iterative method instead, and even in 2D modern iterative methods like PCG [20] and GMRES [31] are faster.

The classical iterative idea is to replace the matrix A by another matrix M where solving $Mu = \ldots$ is simple. M can be, for instance, the diagonal part of A. First write $Au = b$ as

$$Mu = (M - A)u + b \tag{7.5}$$

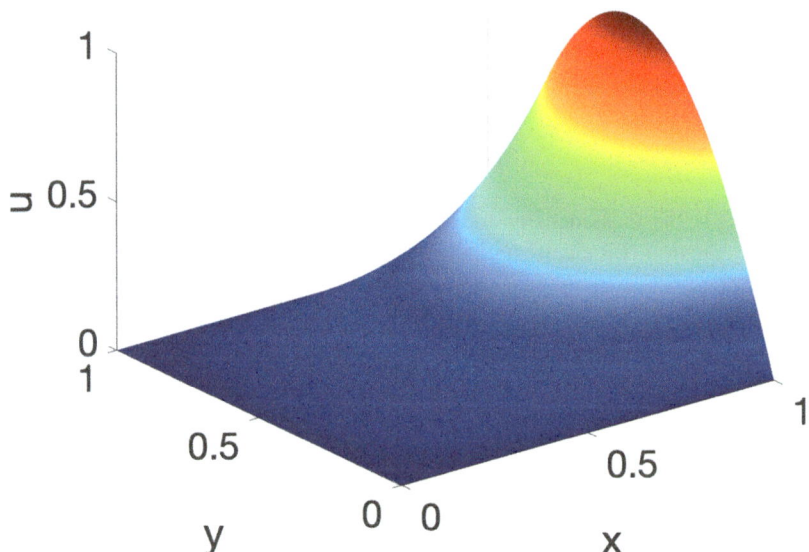

Fig. 7.1 Numerical solution to the model Laplace problem using any of **laplace0.m–laplace6.m** on a 500$\Delta x \times$ 500Δy grid with homogeneous Dirichlet boundary conditions except $u(x = 1, y) = 4y(1 - y)$

and then evaluate u on the right-hand side in terms of a previous guess $u^{(k)}$ to obtain the new improved guess $u^{(k+1)}$:

$$Mu^{(k+1)} = (M - A)u^{(k)} + b, \quad k = 0, 1, 2, \ldots \tag{7.6}$$

Split the matrix $A = L + D + U$, where D is the diagonal part of A, L is the strictly lower triangular part of A, and U is the strictly upper triangular part of A. If $M = D$, we have the *Jacobi* iterative method. If $M = L + D$, we have the *Gauss-Seidel* iterative method. (Alternatively, we could take $M = D + U$.)

A major advance in the efficiency of iterative methods began with *SOR* (successive over-relaxation), presented in Young's PhD thesis in 1950 (see [41]). Here $M = \frac{D}{\omega} + L$, where $0 < \omega < 2$ is a relaxation factor. For over-relaxation, $1 < \omega < 2$, and for Gauss-Seidel $\omega = 1$.

The modern approach to solving $Au = b$ is to iteratively seek the minimum of $P(u) = \frac{1}{2}u^T Au - u^T b$ where $Au = b$. Minimization methods like PCG and GMRES are even faster and more robust than SOR. PCG is recommended for the Laplace and Poisson equations with constant permittivity, which give rise to symmetric positive definite linear systems, while GMRES can be used to advantage for more general elliptic PDEs that produce indefinite asymmetric linear systems—as when solving the nonlinear Laplace or Poisson equation.

For another general introduction to solving elliptic PDEs, see Strang's *Computational Science and Engineering* [36]. See Trefethen and Bau's *Numerical Linear Algebra* [39] for a comprehensive treatment of PCG and GMRES.

7.1 Laplace's and Poisson's Equations

We will develop numerical methods first for the 2D Laplace equation (7.4) and later make some remarks about discretizing the 3D Laplace equation and the 2D and 3D Poisson equation. At the end of the chapter, we will discuss the nonlinear Laplace and Poisson equations.

Figure 7.1 illustrates a model 2D Laplace problem with Dirichlet boundary conditions:

$$u(x = 0, y) = 0, \quad u(x = 1, y) = 4y(1 - y),$$
$$u(x, y = 0) = 0, \quad u(x, y = 1) = 0. \tag{7.7}$$

To discretize Laplace's equation on an $N\Delta x \times N\Delta y$ grid with $\Delta x = \Delta y \equiv h$, set $u_{ij} \approx u(x_i, y_j)$ and use second-order accurate central differences:

$$(u_{i+1,j} + u_{i-1,j} + u_{i,j+1} + u_{i,j-1} - 4u_{ij})/h^2 = 0 \tag{7.8}$$

where $i, j = 1, 2, \ldots, N - 1$. Then, multiplying by h^2,

$$u_{i+1,j} + u_{i-1,j} + u_{i,j+1} + u_{i,j-1} - 4u_{ij} = 0. \tag{7.9}$$

Equation (7.9) can be converted to a matrix-vector equation $Au = b$ or solved as it stands by an iterative method

$$u_{ij} \leftarrow \frac{1}{4}(u_{i+1,j} + u_{i-1,j} + u_{i,j+1} + u_{i,j-1}). \tag{7.10}$$

To write the matrix-vector equation $Au = b$, refer to the five-star stencil in Fig. 7.2 for a $4\Delta x \times 4\Delta y$ example ($N = 4$):

$$Au = \begin{bmatrix} -4 & 1 & 0 & 1 & 0 & 0 & 0 & 0 & 0 \\ 1 & -4 & 1 & 0 & 1 & 0 & 0 & 0 & 0 \\ 0 & 1 & -4 & 0 & 0 & 1 & 0 & 0 & 0 \\ 1 & 0 & 0 & -4 & 1 & 0 & 1 & 0 & 0 \\ 0 & 1 & 0 & 1 & -4 & 1 & 0 & 1 & 0 \\ 0 & 0 & 1 & 0 & 1 & -4 & 0 & 0 & 1 \\ 0 & 0 & 0 & 1 & 0 & 0 & -4 & 1 & 0 \\ 0 & 0 & 0 & 0 & 1 & 0 & 1 & -4 & 1 \\ 0 & 0 & 0 & 0 & 0 & 1 & 0 & 1 & -4 \end{bmatrix} \begin{bmatrix} u_0 \\ u_1 \\ u_2 \\ u_3 \\ u_4 \\ u_5 \\ u_6 \\ u_7 \\ u_8 \end{bmatrix} = b = \begin{bmatrix} -b_0 - b_{11} \\ -b_1 \\ -b_2 - b_3 \\ -b_{10} \\ 0 \\ -b_4 \\ -b_8 - b_9 \\ -b_7 \\ -b_6 - b_5 \end{bmatrix}$$
(7.11)

where b contains only boundary values.

For an $N\Delta x \times N\Delta y$ grid with Dirichlet boundary conditions, A is $(N-1)^2 \times (N-1)^2$, and there are $N-3$ zeros between the diagonals and the fringes. The bandwidth $w = N - 1$.

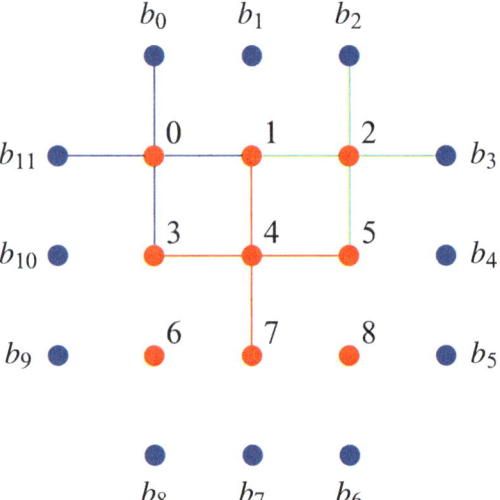

Fig. 7.2 Five-star stencil for the 2D Laplace and Poisson equations

7.1 Laplace's and Poisson's Equations

Table 7.1 1D, 2D, and 3D Laplacian matrices, with memory and complexity estimates for a banded matrix direct solve using Gaussian elimination. The Laplacian matrix is $n \times n$ in the large N limit, with bandwidth w. The memory required for Gaussian elimination due to fill-in is $\sim nw$. In 3D with $N = 100$, Gaussian elimination requires ~ 80 GB of memory with 8-byte doubles, while for $N = 500$, Gaussian elimination requires ~ 250 TB of memory, which is prohibitive. The complexity (operation count)—measured in flops—scales $\sim w^2 n$. See, for example, [35]

Dimension	Grid	n	Bands	w	Memory	Complexity
1D	N	N	3	1	$2N$	$5N$
2D	$N \times N$	N^2	5	N	N^3	N^4
3D	$N \times N \times N$	N^3	7	N^2	N^5	N^7

Table 7.1 presents the number of unknowns, bands, bandwidths, memory usages, and complexities for a banded matrix direct solve (based on Gaussian elimination) for the Laplace problem in 1D, 2D, and 3D. A 2D banded matrix direct solve for the model Laplace problem is provided in **laplace0.m**. The matrix A is constructed with the MATLAB function `-delsq`, and $Au = b$ is solved in MATLAB with the backslash operator.

The matrix-vector equation $Au = b$ can be solved either directly or iteratively. Consider an N^d rectangular grid in d dimensions for the Laplace/Poisson equation. Banded matrix direct methods[1] require

$$N_{direct} \sim w^2 N^d \sim N^{3d-2} \tag{7.12}$$

flops, where $w \sim N^{d-1}$ is the matrix bandwidth, while a modern iterative method like PCG or GMRES requires

$$N_{iter} \sim N^{d+1} \tag{7.13}$$

flops (see, for example, Chap. 4 of [26]). In 1D, the banded matrix direct method is faster, but in 2D and 3D modern iterative methods are faster. In 3D, not only are modern iterative methods much faster than banded/sparse matrix direct solvers, but iterative methods are required due to memory usage.

In 2D, sparse matrix direct methods [16] like nested dissection or minimal degree ordering can be an order of magnitude faster than banded matrix direct methods for the Laplace/Poisson equation, with $N_{sparse} \sim N^3$ flops. The MATLAB backslash operator in `u = A\b` automatically chooses the fastest among these methods, based on the structure of the matrix A.

---

[1] The fast Fourier transform (FFT) method (see [35], e.g.) can be used for the Laplace/Poisson equation on rectangular grids with $N_{FFT} \sim N^2 \ln(N)$ flops in 2D, but does not generalize to complicated geometries or to complicated spatial derivatives, like those that occur in the nonlinear Laplace/Poisson equation or in steady-state fluid dynamics.

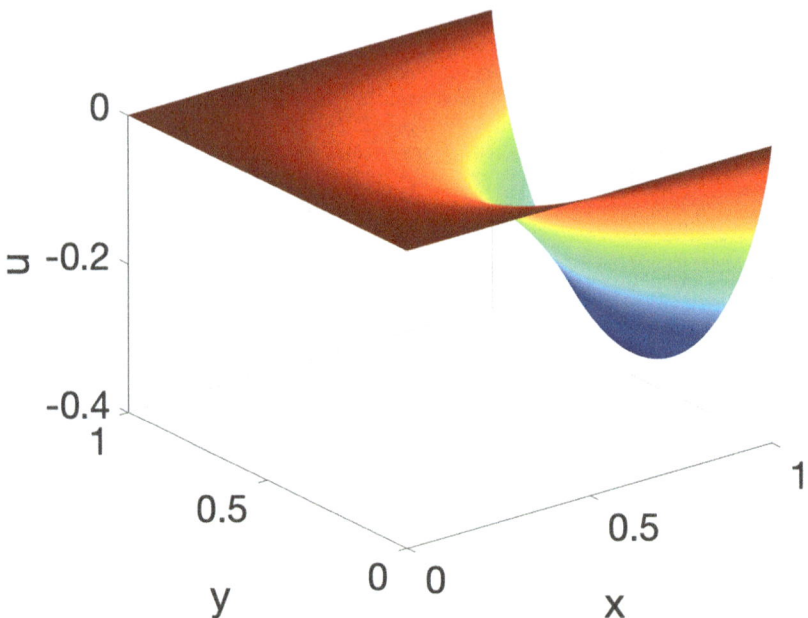

Fig. 7.3 Numerical solution to the model Laplace problem using **laplaceNBC.m** on a $500\Delta x \times 500\Delta y$ grid with homogeneous Dirichlet boundary conditions on three sides and Neumann boundary condition $u_x(x=1, y) = -1$ on the fourth

A 2D Laplace problem with a Neumann boundary condition $u_x(x=1, y) = -1$ (see Sect. 6.10.2) on one side and homogeneous Dirichlet boundary conditions on the other three sides is illustrated in Fig. 7.3.

7.2 Fourier Solution for the Model Laplace Problem

Before discussing iterative methods for Laplace's equation, we first derive the Fourier solution to the model Laplace problem (7.4) and (7.7), which can be used to validate our computational solutions.

The Fourier solution is obtained by making a Fourier sine expansion in y

$$u(x, y) = \sum_{n=1}^{\infty} a_n(x) \sin(n\pi y) \qquad (7.14)$$

since it automatically satisfies the boundary conditions $u(x, y=0) = 0 = u(x, y=1)$. Plugging the Fourier expansion into Laplace's equation (7.4) and using orthogonality of the sines (see Appendix A.4), we have

$$\sum_{n=1}^{\infty} \frac{d^2 a_n}{dx^2} \sin(n\pi y) - \sum_{n=1}^{\infty} n^2 \pi^2 a_n \sin(n\pi y) = 0 \tag{7.15}$$

$$\frac{d^2 a_n}{dx^2} = n^2 \pi^2 a_n, \quad a_n(x) = \alpha_n \sinh(n\pi x) + \beta_n \cosh(n\pi x). \tag{7.16}$$

The boundary condition $u(x = 0, y) = 0$ implies $\beta_n = 0$, and the inhomogeneous Dirichlet boundary condition implies

$$u(x = 1, y) = \sum_{n=1}^{\infty} \alpha_n \sinh(n\pi) \sin(n\pi y) = 4y(1-y) \tag{7.17}$$

or integrating (7.17) $\int_0^1 dy \, \sin(m\pi y)$:

$$\alpha_n = \frac{32}{(n\pi)^3 \sinh(n\pi)}, \quad n = 1, 3, 5, \ldots \tag{7.18}$$

for n odd and $\alpha_n = 0$ for n even. Therefore,

$$u(x, y) = \sum_{n=1}^{\infty} \frac{32}{((2n-1)\pi)^3} \frac{\sinh((2n-1)\pi x)}{\sinh((2n-1)\pi)} \sin((2n-1)\pi y). \tag{7.19}$$

On the computer, $u(x, t)$ is approximated by

$$u_{N_F}(x, y) = \sum_{n=1}^{N_F} \frac{32}{((2n-1)\pi)^3} \frac{\sinh((2n-1)\pi x)}{\sinh((2n-1)\pi)} \sin((2n-1)\pi y) \tag{7.20}$$

where N_F is the number of Fourier modes.

Figure 7.4 demonstrates that the SOR solver in **laplace3.m** for the model Laplace problem yields second-order accuracy by comparing the computed solution with the "exact" Fourier solution on various spatial grids. Similar results are obtained for the other Laplace solvers as well.

7.3 The Classical Iterative Idea

The basic iterative idea is to replace the matrix A in $Au = b$ by another matrix M, where solving $Mu = \ldots$ is simple:

$$Mu = (M - A)u + b. \tag{7.21}$$

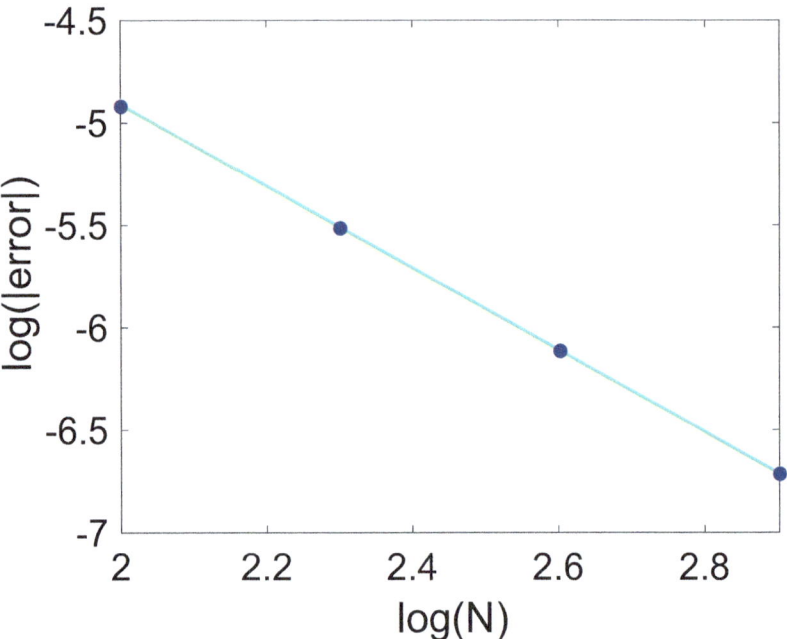

Fig. 7.4 Numerical errors $\log_{10}(\|u_{exact} - u_{comp}\|_1)$ vs. $\log_{10}(N)$ for the model Laplace problem comparing the SOR solution from **laplace3.m** with the "exact" Fourier solution on $N \times N$ grids with $N = 100, 200, 400,$ and 800. The *straight line* is the theoretical value with log-log slope of -2

Now evaluate u iteratively:

$$Mu^{(k+1)} = (M - A)u^{(k)} + b, \quad k = 0, 1, 2, \ldots \qquad (7.22)$$

The error $e^{(k)} = u^{(k)} - u$ in solving the iterative Equation (7.22) satisfies

$$Me^{(k+1)} = (M - A)e^{(k)}, \quad e^{(k+1)} = Be^{(k)} \qquad (7.23)$$

where the iteration matrix $B = M^{-1}(M - A)$. As $k \to \infty$,

$$\left\|e^{(k)}\right\| = \left\|B^k e^{(0)}\right\| \sim \rho^k \to 0 \quad \text{iff} \quad \rho < 1 \qquad (7.24)$$

where the spectral radius ρ of B is

$$\rho = \max_j |\lambda_j| \qquad (7.25)$$

with λ_j the eigenvalues of B.

7.4 Classical Iterative Methods for Laplace's Equation

Table 7.2 Number of iterative sweeps for the model Laplace problem on three $N\Delta x \times N\Delta y$ grids with $\Delta x = \Delta y \equiv h$. For convergence of the iterative methods, $\epsilon = 10^{-5}h^2$

Solver	100 × 100	200 × 200	400 × 400
Jacobi	35,529	147,753	613,505
Gauss-Seidel	17,743	73,740	306,105
SOR	400	806	1645

Thus we need to know the spectral radius of the iteration matrix B to evaluate convergence of the iterative method. For the model Laplace problem on an $N\Delta x \times N\Delta y$ grid with $\Delta x = \Delta y \equiv h$ and $h = 1/N$, the Jacobi, Gauss-Seidel, and SOR spectral radii are (see Problems 7.4 and 7.5)

$$\rho_J = \cos(\pi h) \approx 1 - \frac{\pi^2 h^2}{2}, \quad \rho_{GS} = \rho_J^2 = \cos^2(\pi h) \approx 1 - \pi^2 h^2 \quad (7.26)$$

$$\rho_{SOR} = \frac{1 - \sin \pi h}{1 + \sin \pi h} \approx 1 - 2\pi h. \quad (7.27)$$

7.4 Classical Iterative Methods for Laplace's Equation

In this section, we will develop the classical Jacobi, Gauss-Seidel, and SOR iterative methods for Laplace's equation. We will write out the iterative methods both as pointwise formulas (most useful for programming) and in matrix-vector form (most useful for theoretical analysis). Then in the next section, we will develop the general theory of the classical iterative methods.

Table 7.2 presents the number of iterative sweeps for Jacobi, Gauss-Seidel, and SOR for the model Laplace problem on three $N\Delta x \times N\Delta y$ grids with $\Delta x = \Delta y \equiv h$. Note that the number of Gauss-Seidel iterations is approximately half the number of Jacobi iterations and that the number of SOR iterations is roughly $1/N$ times the number of Jacobi iterations, as predicted by theory.

For computing the iterative solution to Laplace's equation, the best approach is to write the Jacobi, Gauss-Seidel, and SOR methods in terms of the residual defined (at iteration $k = 0, 1, 2, \ldots$) by

$$r_{ij}^{(k)} = -4u_{ij}^{(k)} + u_{i+1,j}^{(k)} + u_{i-1,j}^{(k)} + u_{i,j+1}^{(k)} + u_{i,j-1}^{(k)}. \quad (7.28)$$

In matrix form, the residual (at iteration k) is

$$r^{(k)} = Au^{(k)} - b. \quad (7.29)$$

The Gauss-Seidel and SOR methods can be expressed most simply by using the *current* residual $\tilde{r}_{ij}$, where some nearest neighbor values of the solution have already been updated. For example, updating along rows from left to right and top to bottom:

$$\tilde{r}_{ij} = -4u_{ij}^{(k)} + u_{i+1,j}^{(k)} + u_{i-1,j}^{(k+1)} + u_{i,j+1}^{(k+1)} + u_{i,j-1}^{(k)} \tag{7.30}$$

or in matrix form:

$$\tilde{r} = Lu^{(k+1)} + (D+U)u^{(k)} - b. \tag{7.31}$$

7.4.1 Jacobi Iteration

Write $A = L + D + U$ as the sum of its diagonal part and strictly lower and strictly upper triangular parts.

The matrix $M_J = D$ for Jacobi iteration is

$$M_J = \begin{bmatrix} -4 & 0 & 0 & 0 & 0 & 0 & 0 & 0 & 0 \\ 0 & -4 & 0 & 0 & 0 & 0 & 0 & 0 & 0 \\ 0 & 0 & -4 & 0 & 0 & 0 & 0 & 0 & 0 \\ 0 & 0 & 0 & -4 & 0 & 0 & 0 & 0 & 0 \\ 0 & 0 & 0 & 0 & -4 & 0 & 0 & 0 & 0 \\ 0 & 0 & 0 & 0 & 0 & -4 & 0 & 0 & 0 \\ 0 & 0 & 0 & 0 & 0 & 0 & -4 & 0 & 0 \\ 0 & 0 & 0 & 0 & 0 & 0 & 0 & -4 & 0 \\ 0 & 0 & 0 & 0 & 0 & 0 & 0 & 0 & -4 \end{bmatrix}. \tag{7.32}$$

Jacobi's method is

$$u_{ij}^{(k+1)} = \frac{1}{4}\left(u_{i+1,j}^{(k)} + u_{i-1,j}^{(k)} + u_{i,j+1}^{(k)} + u_{i,j-1}^{(k)}\right) = u_{ij}^{(k)} + \frac{1}{4}r_{ij}^{(k)} \quad \text{(Jacobi)} \tag{7.33}$$

or in matrix form

$$Du^{(k+1)} = -(L+U)u^{(k)} + b = Du^{(k)} - r^{(k)} \tag{7.34}$$

$$B_J = -D^{-1}(L+U) \tag{7.35}$$

$$u^{(k+1)} = u^{(k)} - D^{-1}r^{(k)} \quad \text{(Jacobi)}. \tag{7.36}$$

Note that $D^{-1} = -\frac{1}{4}I$.

Our MATLAB code **laplace1.m** solves the 2D model Laplace problem with Jacobi iteration using array operations.

7.4 Classical Iterative Methods for Laplace's Equation

7.4.2 Gauss-Seidel Iteration

The matrix $M_{GS} = D + L$ for Gauss-Seidel iteration is

$$M_{GS} = \begin{bmatrix} -4 & 0 & 0 & 0 & 0 & 0 & 0 & 0 & 0 \\ 1 & -4 & 0 & 0 & 0 & 0 & 0 & 0 & 0 \\ 0 & 1 & -4 & 0 & 0 & 0 & 0 & 0 & 0 \\ 1 & 0 & 0 & -4 & 0 & 0 & 0 & 0 & 0 \\ 0 & 1 & 0 & 1 & -4 & 0 & 0 & 0 & 0 \\ 0 & 0 & 1 & 0 & 1 & -4 & 0 & 0 & 0 \\ 0 & 0 & 0 & 1 & 0 & 0 & -4 & 0 & 0 \\ 0 & 0 & 0 & 0 & 1 & 0 & 1 & -4 & 0 \\ 0 & 0 & 0 & 0 & 0 & 1 & 0 & 1 & -4 \end{bmatrix}. \quad (7.37)$$

Gauss-Seidel can be written as

$$u_{ij}^{(k+1)} = u_{ij}^{(k)} + \frac{1}{4}\tilde{r}_{ij} \quad \text{(Gauss-Seidel)} \quad (7.38)$$

or in matrix form as

$$(D+L)u^{(k+1)} = -Uu^{(k)} + b \quad (7.39)$$

$$B_{GS} = -(L+D)^{-1}U \quad (7.40)$$

$$Du^{(k+1)} = Du^{(k)} - \left(Lu^{(k+1)} + Du^{(k)} + Uu^{(k)} - b\right) = Du^{(k)} - \tilde{r} \quad (7.41)$$

$$u^{(k+1)} = u^{(k)} - D^{-1}\tilde{r} \quad \text{(Gauss-Seidel)}. \quad (7.42)$$

Gauss-Seidel iteration for the 2D model Laplace problem is implemented in our MATLAB code **laplace2.m** with for-loops (more natural here than with array operations).

Parallelizing for-loops (in C and C++) or do-loops (in Fortran) with OpenMP is straightforward.

7.4.3 SOR Iteration

The matrix $M_{SOR} = \frac{D}{\omega} + L$ for SOR iteration is

$$M_{SOR} = \begin{bmatrix} -4/\omega & 0 & 0 & 0 & 0 & 0 & 0 & 0 & 0 \\ 1 & -4/\omega & 0 & 0 & 0 & 0 & 0 & 0 & 0 \\ 0 & 1 & -4/\omega & 0 & 0 & 0 & 0 & 0 & 0 \\ 1 & 0 & 0 & -4/\omega & 0 & 0 & 0 & 0 & 0 \\ 0 & 1 & 0 & 1 & -4/\omega & 0 & 0 & 0 & 0 \\ 0 & 0 & 1 & 0 & 1 & -4/\omega & 0 & 0 & 0 \\ 0 & 0 & 0 & 1 & 0 & 0 & -4/\omega & 0 & 0 \\ 0 & 0 & 0 & 0 & 1 & 0 & 1 & -4/\omega & 0 \\ 0 & 0 & 0 & 0 & 0 & 1 & 0 & 1 & -4/\omega \end{bmatrix}.$$
(7.43)

The Gauss-Seidel approximation $u^{(k+1)}$ tends to stay on the same side of the root of $Au - b = 0$ (in the n-dimensional space). SOR is intuitively derived by simply overcorrecting Gauss-Seidel:

$$u_{ij}^{(k+1)} = u_{ij}^{(k)} + \frac{\omega}{4} \tilde{r}_{ij}, \quad 1 < \omega < 2 \quad \text{(SOR)} \tag{7.44}$$

or in matrix form (and working backward):

$$u^{(k+1)} = u^{(k)} - \omega D^{-1} \tilde{r} \quad \text{(SOR)} \tag{7.45}$$

$$\frac{D}{\omega} u^{(k+1)} = \frac{D}{\omega} u^{(k)} - \left(L u^{(k+1)} + D u^{(k)} + U u^{(k)} - b \right) \tag{7.46}$$

$$\left(\frac{D}{\omega} + L \right) u^{(k+1)} = \left(\frac{D}{\omega} - D - U \right) u^{(k)} + b \tag{7.47}$$

$$B_{SOR} = \left(\frac{D}{\omega} + L \right)^{-1} \left(\frac{D}{\omega} - D - U \right) = (D + \omega L)^{-1} ((1 - \omega) D - \omega U). \tag{7.48}$$

Similarly, *SUR* (successive under-relaxation)—necessary, for example, for steady-state hyperbolic problems—has $0 < \omega < 1$. The relaxation factor $0 < \omega < 2$ in SOR/SUR is adjusted to make the spectral radius ρ as small as possible (see Fig. 7.5).

Our MATLAB code **laplace3.m** implements SOR for the 2D model Laplace problem with for-loops.

To prove $0 < \omega < 2$, first observe that $B_{SOR/SUR}$ is the product of triangular matrices. Recall if T is a triangular matrix, $\det\{T\} = \prod_i T_{ii}$ (the product of the diagonal elements), and for any matrices M and N: $\det\{MN\} = \det\{M\} \det\{N\}$ and $\det\{M^{-1}\} = 1/\det\{M\}$.

Then for SOR/SUR,

$$\det\{B\} = (1 - \omega)^n \tag{7.49}$$

7.4 Classical Iterative Methods for Laplace's Equation

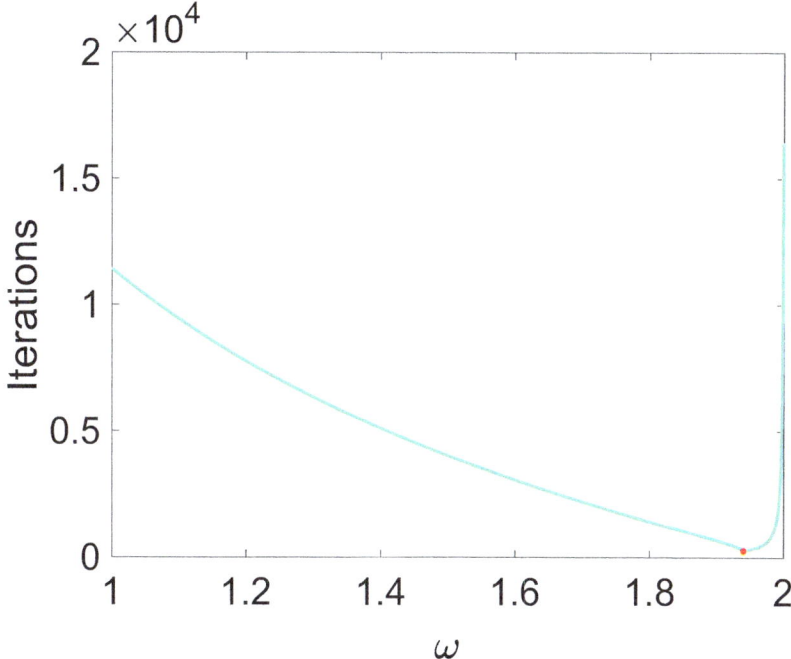

Fig. 7.5 Number of sweeps required for convergence of SOR vs. ω on a 100×100 grid for the model Laplace problem, with $\epsilon = 10^{-6}$ for convergence. The *dot* is for $\omega_{opt} = 1.93968$, which requires only 306 iterations for convergence. By way of contrast, Gauss-Seidel ($\omega = 1$) requires 11,448 iterations for convergence

and

$$|1 - \omega|^n = |\det\{B\}| = |\lambda_1||\lambda_2| \cdots |\lambda_n| \leq \rho^n < 1 \tag{7.50}$$

for convergence, which implies that

$$|1 - \omega| < 1 \quad \text{iff} \quad 0 < \omega < 2. \tag{7.51}$$

For the optimal ω_{opt}, $\rho = |1 - \omega_{opt}|$: SOR has $1 < \omega < 2$ and $\rho_{SOR} = \omega_{opt} - 1$; SUR has $0 < \omega < 1$ and $\rho_{SUR} = 1 - \omega_{opt}$. In both cases, the eigenvalues of B have moved onto a circle of radius ρ_{opt} in the complex plane.

By dynamically adjusting ω, the Chebyshev SOR method insures that the norm of the error always decreases. Make two half sweeps on even/odd (black/white) meshes. Odd (even) points depend only on even (odd) mesh values. Define (μ is the Jacobi spectral radius):

$$\omega^{(0)} = 1 \tag{7.52}$$

$$\omega^{(\frac{1}{2})} = \frac{1}{1 - \frac{\mu^2}{2}} \tag{7.53}$$

$$\omega^{(k+\frac{1}{2})} = \frac{1}{1 - \frac{\mu^2 \omega^{(k)}}{4}}, \quad k = \frac{1}{2}, 1, \frac{3}{2}, \ldots \tag{7.54}$$

Then $\omega^{(\infty)} = \omega_{opt}$, and it can be proved that the norm of the error always decreases.

On parallel computers, chaotic relaxation may be used with Gauss-Seidel or SOR/SUR by updating solution values on multiple processors asynchronously. As long as every solution value is updated from time to time, the method will converge, giving linear speedup in the number of processors.

7.5 Theory of Classical Iterative Methods

As is traditional in discussing the theory of iterative methods, we rewrite the linear equation to be solved as $Ax = b$.

To briefly review: To solve $Ax = b$, write

$$Mx^{(k+1)} = (M - A)x^{(k)} + b, \quad k = 0, 1, 2, \ldots \tag{7.55}$$

Then the error $e^{(k)} = x^{(k)} - x$ satisfies

$$Me^{(k+1)} = (M - A)e^{(k)}, \quad e^{(k+1)} = Be^{(k)} \tag{7.56}$$

where the iteration matrix $B = M^{-1}(M - A)$. Now as $k \to \infty$,

$$\left\|e^{(k)}\right\| = \left\|B^k e^{(0)}\right\| \sim \rho^k \to 0 \quad \text{iff} \quad \rho < 1 \tag{7.57}$$

where ρ is the spectral radius of B.

Two important questions concern *convergence:* (i) If the iterative method converges, does it converge to the exact solution? If $x^{(k)} \to x^*$, then

$$Mx^* = (M - A)x^* + b \tag{7.58}$$

and $Ax^* = b$, which implies $x^* = x$. (ii) If the iterative method converges, how fast does it converge? The rate of convergence is governed by powers of $B = M^{-1}(M - A)$. Since $e^{(k)} = B^k e^{(0)}$, $e^{(k)} \to 0$ and $x^{(k)} \to x$ (convergence) if $B^k \to 0$ (stability), which incidentally proves the Equivalence Theorem in this context.

Theorem $B^k \to 0$ *iff every eigenvalue of B satisfies $|\lambda_j| < 1$. The rate of convergence is governed by the spectral radius ρ of B:*

7.5 Theory of Classical Iterative Methods

$$\rho = \max_j |\lambda_j|. \tag{7.59}$$

Proof (Using eigenvectors) Expand the initial error in terms of the eigenvectors of B (assuming B is a *normal* matrix[2]):

$$e^{(0)} = c_1 v_1 + c_2 v_2 + \cdots + c_n v_n \tag{7.60}$$

where $Bv_j = \lambda_j v_j$. Then as $k \to \infty$, the error at iteration k is

$$e^{(k)} = B^k e^{(0)} = c_1 \lambda_1^k v_1 + c_2 \lambda_2^k v_2 + \cdots + c_n \lambda_n^k v_n \to 0 \quad \text{iff} \quad |\lambda_j| < 1 \tag{7.61}$$

for $j = 1, 2, \ldots, n$, and

$$\left\| e^{(k)} \right\| \sim \rho^k \tag{7.62}$$

as $k \to \infty$.

Proof (In matrix form) Assuming a complete set of eigenvectors, write $B = S \Lambda S^{-1}$, where $S = [v_1 \ v_2 \ \cdots \ v_n]$ is the eigenvector matrix of B and

$$\Lambda = \operatorname{diag}\{\lambda_1, \lambda_2, \ldots, \lambda_n\} \equiv \begin{bmatrix} \lambda_1 & & & \\ & \lambda_2 & & \\ & & \ddots & \\ & & & \lambda_n \end{bmatrix} \tag{7.63}$$

is the eigenvalue matrix of B (white space denotes 0s). Then

$$B^k = \left(S \Lambda S^{-1}\right)\left(S \Lambda S^{-1}\right) \cdots \left(S \Lambda S^{-1}\right) = S \Lambda^k S^{-1} \to 0 \quad \text{iff} \quad |\lambda_j| < 1 \tag{7.64}$$

for $j = 1, 2, \ldots, n$.

If B does not have a complete set of eigenvectors, the matrix form of the proof involves the Jordan normal form J of B instead of Λ, where

$$J = \begin{bmatrix} J_1 & & & \\ & J_2 & & \\ & & \ddots & \\ & & & J_m \end{bmatrix}, \quad J_j = \begin{bmatrix} \lambda_j & 1 & & \\ & \lambda_j & \ddots & \\ & & \ddots & 1 \\ & & & \lambda_j \end{bmatrix} \quad \text{for } j = 1, 2, \ldots, m. \tag{7.65}$$

---

[2] A normal matrix A has a complete set of eigenvectors. The test for a normal matrix is whether $A^\dagger A = A A^\dagger$, with the Hermitian conjugate $A^\dagger = (A^*)^T$. For example, real symmetric matrices are normal.

For a wide class of finite difference matrices, Young's formula relates the eigenvalues μ of Jacobi and the eigenvalues λ of SOR:

$$(\lambda + \omega - 1)^2 = \lambda \omega^2 \mu^2. \tag{7.66}$$

Minimizing $\rho_{SOR} = \max\{|\lambda|\} = \omega - 1$ (using the quadratic formula) gives

$$\omega_{opt} = \frac{2\left(1 - \sqrt{1 - \rho_J^2}\right)}{\rho_J^2}, \quad \rho_{SOR} = \omega_{opt} - 1. \tag{7.67}$$

For the model Laplace problem on an $N\Delta x \times N\Delta y$ grid with $\Delta x = \Delta y \equiv h$ and $h = 1/N$,

$$\rho_J = \cos(\pi h), \quad \rho_{GS} = \rho_J^2, \quad \rho_{SOR} = \frac{1 - \sin \pi h}{1 + \sin \pi h}. \tag{7.68}$$

To show the Jacobi spectral radius $\rho_J = \cos(\pi h)$ for Laplace's equation on the unit square with second-order accurate central differences, first consider a 1D problem. In 1D, set $A = \text{tridiag}[1, -2, 1]$ where A is $n \times n$ for n interior grid points plus Dirichlet boundary conditions. Then the iteration matrix $B = \frac{1}{2}\text{tridiag}[-1, 0, -1]$. Show $Bv = \cos(\pi h) v$ where the 1D eigenvector

$$v = [\sin(\pi h), \sin(2\pi h), \cdots, \sin(n\pi h)], \quad h = \frac{1}{n+1} \tag{7.69}$$

(see Problem 7.4). The other eigenvectors of B replace π with $2\pi, 3\pi, \ldots, n\pi$ in v, with eigenvalues $\cos(2\pi h), \cos(3\pi h), \ldots, \cos(n\pi h)$.

In 2D, the eigenvector

$$v = [\sin(j\pi h) \sin(k\pi h)]$$

$$= [\sin(\pi h) \sin(\pi h), \sin(\pi h) \sin(2\pi h), \cdots, \sin(\pi h) \sin(n\pi h),$$

$$\sin(2\pi h) \sin(\pi h), \cdots, \sin(2\pi h) \sin(n\pi h), \cdots, \sin(n\pi h) \sin(n\pi h)] \tag{7.70}$$

with the same eigenvalue, so $\rho_J = \cos(\pi h)$.

Using ρ_J in Young's formula yields the SOR spectral radius

$$\rho_{SOR} = \frac{1 - \sin \pi h}{1 + \sin \pi h}. \tag{7.71}$$

7.6 Conjugate Gradients

The modern approach to solving $Ax = b$ is to iteratively seek the minimum of

$$P(x) = \frac{1}{2}x^T Ax - x^T b \tag{7.72}$$

where $Ax = b$. Minimization methods like PCG (preconditioned conjugate gradients) and GMRES (generalized minimal residuals) are even faster and more robust than SOR. We start with the method of steepest descent, which is not guaranteed to reach the minimum of $P(x)$ in a finite number of steps; then we develop CG and PCG, which are guaranteed (in exact arithmetic) to reach the minimum in n steps for an $n \times n$ matrix A (see Fig. 7.6).

Note that in practice, we almost always stop the iterative procedure before n steps for two reasons: first, we are working in finite precision arithmetic, so after a certain point, successive iterates may not be more accurate; and second, all we need is an approximate solution to $Ax = b$, which may be obtained in far fewer than n steps.

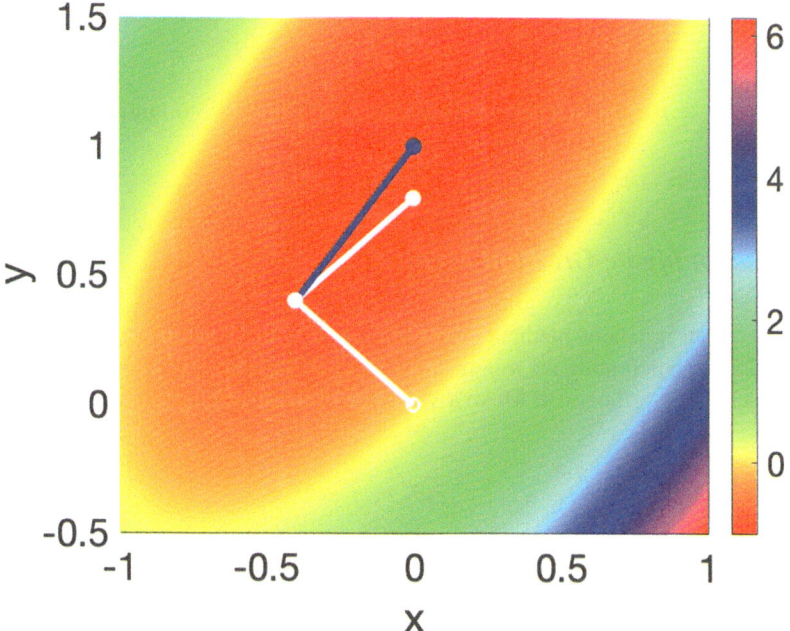

Fig. 7.6 Two steps of steepest descent (white lines) vs. two steps of conjugate gradients (*dark line*) for $P(x, y) = 2x^2 - 2xy + y^2 + 2x - 2y$. The first step of CG is the same as steepest descent. The starting point $(0, 0)$ is denoted by an *open white circle*. The exact answer $(0, 1)$ (*dark dot*) is obtained by CG in two steps for two unknowns, while steepest descent approaches but never reaches the exact answer in a finite number of steps

Preconditioning the matrix A will allow PCG to converge to a given precision in fewer steps than CG.

The material in Sects. 7.6.1–7.6.4 is based on Strang's treatment in [35, 36].

It is not hard to program the CG and PCG algorithms (see our **laplace4.m** and **laplace5.m** MATLAB codes), but it's best to use numerical library codes for PCG and GMRES (see our **laplace6.m**).

7.6.1 Method of Steepest Descent

The method of steepest descent tries to find the minimum of $P(x) = \frac{1}{2}x^T A x - x^T b$ by using gradient information. It is not guaranteed to reach the minimum though in a finite number of steps.

The minimum of P solves $Ax = b$:

$$-\nabla P = 0 = -Ax + b \equiv r \quad \text{iff} \quad Ax = b \tag{7.73}$$

where r is the residual. Make a sequence of guesses $x^{(0)}, x^{(1)}, x^{(2)}, \ldots$ to x, where

$$x^{(j)} = x^{(j-1)} - \alpha \, \nabla P|_{x^{(j-1)}}, \quad j = 1, 2, \ldots \tag{7.74}$$

Here $-\nabla P = r$ is the steepest descent direction. Choose α to minimize P in that direction.

7.6.2 Conjugate Gradient Method

The conjugate gradient method solves $Ax = b$ with A symmetric and positive definite[3] by again finding the minimum of $P(x) = \frac{1}{2}x^T Ax - x^T b$, but now making the directions A-orthogonal: Make a sequence of iterative guesses $x_0, x_1, x_2, \ldots$ to x:

$$x_j = x_{j-1} + \alpha_j d_j, \quad j = 1, 2, \ldots, n \tag{7.75}$$

with d_1 the steepest descent direction and each new direction d_j A-orthogonal to the previous directions:

---

[3] A symmetric matrix is *positive definite* iff $x^T Ax > 0$ for all $x \neq 0$ iff all its eigenvalues are positive iff all its pivots are positive. If A is not symmetric and positive definite, multiply $Ax = b$ by A^T and solve $(A^T A)x = A^T b$.

7.6 Conjugate Gradients

$$x = x_n = \alpha_1 d_1 + \alpha_2 d_2 + \cdots + \alpha_n d_n, \quad d_i^T A d_j = 0 \text{ for } i \neq j, \quad i, j = 1, 2, \ldots, n. \quad (7.76)$$

Then each x_j is the best choice in that

$$x_j = \alpha_1 d_1 + \alpha_2 d_2 + \cdots + \alpha_j d_j \quad (7.77)$$

and the exact solution $x = x_n$. Note that each change $x_{j+1} - x_j$ is *conjugate* (A-orthogonal) to all previous changes:

$$(x_{i+1} - x_i)^T A (x_{j+1} - x_j) = 0, \quad i \neq j. \quad (7.78)$$

To find the α's, use the fact $d_j^T b = d_j^T A x = \alpha_j d_j^T A d_j$. Then

$$\alpha_j = \frac{d_j^T b}{d_j^T A d_j} = \frac{r_{j-1}^T r_{j-1}}{d_j^T A d_j}. \quad (7.79)$$

In implementing the algorithm, start from $x_0 = 0$. Take one step of steepest descent:

$$r_0 = b = -\nabla P = d_1. \quad (7.80)$$

Go in the direction d_1 to $x_1 = \alpha_1 d_1$. For steepest descent, the second direction would be $r_1 = b - A x_1$. Correct to make d_2 A-orthogonal to d_1:

$$d_2 = r_1 + \beta_2 d_1. \quad (7.81)$$

Go to $x_2 = x_1 + \alpha_2 d_2$, etc. The β's are chosen to make the direction d_j conjugate to all previous directions d_i:

$$\beta_j = \frac{r_{j-1}^T r_{j-1}}{r_{j-2}^T r_{j-2}}, \quad j \geq 2, \quad \beta_1 = 0. \quad (7.82)$$

Problem 7.8 proves formulas (7.79) and (7.82).

To treat convergence of the CG method, we will make use of the Cayley-Hamilton Theorem, which states that for an $n \times n$ matrix A,

$$(A - \lambda_1 I)(A - \lambda_2 I) \cdots (A - \lambda_n I) = 0. \quad (7.83)$$

Multiplying the Cayley-Hamilton formula by A^{-1}, we obtain A^{-1} as a linear combination of $I, A, A^2, \ldots, A^{n-1}$. Thus $x = A^{-1} b$ is a linear combination of $b, Ab, A^2 b, \ldots, A^{n-1} b$. The conjugate gradient algorithm constructs x_j as a linear combination of $b, Ab, A^2 b, \ldots, A^{j-1} b$.

Proof Since $x_1 \sim b$, $x_2 \sim Ab$, and r_{j-1}, d_j, and $x_j \sim A^{j-1}b$, x_n is a linear combination of $b, Ab, A^2b, \ldots, A^{n-1}b = x$.

The space

$$\mathcal{K}_j = \text{span}(b, Ab, A^2b, \ldots, A^{j-1}b) \tag{7.84}$$

is called a Krylov space of dimension j.

In conjugate gradients, it can be shown the error at step j is bounded by

$$e_j \leq e_0 \left(\frac{\sqrt{\kappa}-1}{\sqrt{\kappa}+1}\right)^{2j} \tag{7.85}$$

where $\kappa = |\max\{|\lambda|\}/\min\{|\lambda|\}$ is the condition number of A.

7.6.3 CG Algorithm

To solve $Ax = b$ with A symmetric and positive definite (if not, multiply $Ax = b$ by A^T), with residual $r = b - Ax$:

- Initial guess $x_0 = 0$ and initial residual $r_0 = b$ (steepest descent).
- Set $j = 1, 2, \ldots$ and loop until $||r_j|| \leq \epsilon ||r_0||$, with $\epsilon = 10^{-6}$, for example.
  $\beta_j = r_{j-1}^T r_{j-1} / (r_{j-2}^T r_{j-2})$ except $\beta_1 = 0$ to insure $d_j^T A d_i = 0, i \neq j$.
  Direction $d_j = r_{j-1} + \beta_j d_{j-1}$ except $d_1 = r_0$.
  $\alpha_j = r_{j-1}^T r_{j-1} / (d_j^T A d_j)$.
  New guess $x_j = x_{j-1} + \alpha_j d_j$.
  New residual $r_j = r_{j-1} - \alpha_j A d_j$ with $r_j^T r_i = 0, i \neq j$.

In exact arithmetic, $x_n = x$.

7.6.4 PCG Algorithm

To speed up convergence, precondition: $M^{-1}Ax = M^{-1}b$ with M symmetric positive definite. A popular choice for M is the incomplete Cholesky factorization[4] of A.

- Initial guess $x_0 = 0$ and initial residual $r_0 = b$ (steepest descent).
- Set $j = 1, 2, \ldots$ and loop until $||r_j|| \leq \epsilon ||r_0||$, with $\epsilon = 10^{-6}$, for example.
  Solve $M z_{j-1} = r_{j-1}$ for z_{j-1}.

---

[4] The incomplete Cholesky factorization of a symmetric positive definite matrix A is a sparse approximation of the Cholesky factorization $A = LL^\dagger$, where L is a lower triangular matrix.

$\beta_j = z_{j-1}^T r_{j-1}/(z_{j-2}^T r_{j-2})$ except $\beta_1 = 0$ to insure $d_j^T S d_i = 0, i \neq j$ where $S = M^{-\frac{1}{2}} A M^{-\frac{1}{2}}$.

Direction $d_j = z_{j-1} + \beta_j d_{j-1}$ except $d_1 = z_0$.
$\alpha_j = z_{j-1}^T r_{j-1}/(d_j^T A d_j)$.
New guess $x_j = x_{j-1} + \alpha_j d_j$.
New residual $r_j = r_{j-1} - \alpha_j A d_j$ with $z_j^T M z_i = 0, i \neq j$.

7.7 GMRES

The linear Poisson and Laplace equations with constant permittivity ε produce symmetric positive definite linear systems upon discretization, which can be solved efficiently by PCG. For more general elliptic PDEs that produce asymmetric linear systems (like the Jacobian J in solving the nonlinear Laplace equation in Sect. 7.8), GMRES should be used instead of PCG. Like CG and PCG, GMRES is guaranteed (in exact arithmetic) to reach the minimum of $P(x) = \frac{1}{2} x^T A x - x^T b$ in n steps for an $n \times n$ matrix A. However, the complexity of each GMRES iteration j increases $\sim j^2$, so the GMRES method is usually restarted after a certain iteration number $k \ll n$, using x_k now as the initial guess.

In addition, the same remarks about stopping the iterative procedure apply as before, since we are working in finite precision arithmetic, and we need just an approximate solution to $Ax = b$. Preconditioning the matrix A, just as in PCG, may allow GMRES to converge to a given precision in fewer steps.

See **laplace6.m** and **nonlin_laplace.m** for solving the linear and nonlinear Laplace equations using the built-in MATLAB GMRES function, plus Problem 7.10 for numerical experiments.

We briefly describe the GMRES algorithm, following Trefethen and Bau [39]. See [31] for an in-depth treatment.

In minimal residual methods for $Ax = b$, where A is $n \times n$ and non-singular, the approximation x_j, $j = 1, 2, \ldots$, in the Krylov subspace $\mathcal{K}_j$ (7.84) is chosen to minimize the Euclidean norm (2-norm) of the residual $||b - Ax_j||$ by using the least squares method (all the norms in this section are 2-norms). Note $x_0 = 0$ implies $r_0 = b$.

The vectors $b, Ab, \ldots, A^{j-1}b$ tend to become linearly dependent as j increases:

$$A^j b \sim \lambda_{max}^j v_{max} \qquad (7.86)$$

where λ_{max} and v_{max} are the maximum eigenvalue in magnitude and its eigenvector (cf. the power method for calculating λ_{max}). Instead we use Arnoldi iteration to find an orthonormal basis $\{q_1, q_2, \ldots, q_j\}$ for $\mathcal{K}_j$. The vector x_j is now written as $x_j = Q_j y$, where Q_j is the $n \times j$ matrix formed by $[q_1 \; q_2 \; \ldots \; q_j]$. Finding the approximation x_j is transformed to finding the vector y that minimizes the residual.

During the Arnoldi iteration process, the $(j + 1) \times j$ upper Hessenberg[5] matrix $\widetilde{H}_j$ is constructed that satisfies

$$AQ_j = Q_{j+1}\widetilde{H}_j. \tag{7.87}$$

Since the columns of Q_j are orthonormal,

$$||r_j|| = ||Ax_j - b|| = ||AQ_j y - b|| = ||Q_{j+1}\widetilde{H}_j y - b|| \tag{7.88}$$

using the identity above. Multiplying by Q_{j+1} on the left inside the norm does not change the norm:

$$||r_j|| = ||\widetilde{H}_j y - Q_{j+1} b||. \tag{7.89}$$

Finally using $Q_{j+1} b = ||b|| e_1$, where $e_1 = (1, 0, \ldots, 0)^T$, we obtain

$$||r_j|| = ||\widetilde{H}_j y - ||b|| e_1|| \tag{7.90}$$

(see Problem 7.9). Hence, x_j can be found by minimizing the norm of the residual (7.90), a linear least squares[6] problem of size j.

7.7.1 GMRES Algorithm

- Set $q_1 = b/||b||$.
- Set $j = 1, 2, \ldots$ and loop until $||r_j|| \leq \epsilon ||r_0||$, with $\epsilon = 10^{-6}$, for example.
  Calculate q_j with the Arnoldi method (Sect. 7.7.2).
  Using least squares, find y to minimize $||r_j|| = ||\widetilde{H}_j y - ||b|| e_1||$.
  Compute $x_j = Q_j y$.

7.7.2 Arnoldi Iteration Algorithm

Arnoldi iteration is a modified Gram-Schmidt procedure to orthonormalize basis vectors.

- Set $q_1 = b/||b||$.
- For $j = 1, 2, \ldots$
  $v = Aq_j$

---

[5] An upper Hessenberg matrix has zeros below the first subdiagonal.
[6] See [36] and [39] for the least squares algorithm.

For $k = 1, 2, \ldots, j$
$$H_{kj} = q_k^T v$$
$$v = v - H_{kj} q_k$$
$$H_{j+1,j} = \|v\|$$
$$q_{j+1} = v/H_{j+1,j} \, .$$

The $(j+1) \times j$ upper Hessenberg matrix $\tilde{H}_j$ is defined by

$$\tilde{H}_j = \begin{bmatrix} H_{11} & H_{12} & \cdots & H_{1j} \\ H_{21} & H_{22} & \cdots & H_{2j} \\ & \ddots & \ddots & \vdots \\ & & H_{j,j-1} & H_{jj} \\ & & & H_{j+1,j} \end{bmatrix}. \tag{7.91}$$

7.8 Nonlinear Laplace Equation: Newton's Method

As an example of a nonlinear elliptic PDE, we will solve the 2D nonlinear Laplace equation

$$\nabla \cdot (\varepsilon(u) \nabla u) = \frac{\partial}{\partial x}\left(\varepsilon(u)\frac{\partial u}{\partial x}\right) + \frac{\partial}{\partial y}\left(\varepsilon(u)\frac{\partial u}{\partial y}\right) = 0 \tag{7.92}$$

where the permittivity ε is now a function of the potential u. We will take the same Dirichlet boundary conditions

$$u(x = 0, y) = 0, \quad u(x = 1, y) = 4y(1 - y),$$
$$u(x, y = 0) = 0, \quad u(x, y = 1) = 0 \tag{7.93}$$

as in the original model Laplace problem and examine the differences in the solutions. Note that the Jacobian matrix J in the linear solve (7.101) is asymmetric.

7.8.1 Discretization and Newton's Method

Discretizing with second-order accurate central differences, we obtain the discrete equations

$$0 = \frac{1}{\Delta x^2} \left(\varepsilon_{i+\frac{1}{2},j}(u_{i+1,j} - u_{ij}) - \varepsilon_{i-\frac{1}{2},j}(u_{ij} - u_{i-1,j}) \right)$$
$$+ \frac{1}{\Delta y^2} \left(\varepsilon_{i,j+\frac{1}{2}}(u_{i,j+1} - u_{ij}) - \varepsilon_{i,j-\frac{1}{2}}(u_{ij} - u_{i,j-1}) \right) \tag{7.94}$$

where $i = 0, 1, \ldots, N_x$, $j = 0, 1, \ldots, N_y$, and

$$\varepsilon_{ij} = \varepsilon(u_{ij}), \quad \varepsilon_{i\pm\frac{1}{2},j} = \frac{1}{2}\left(\varepsilon_{ij} + \varepsilon_{i\pm1,j}\right), \quad \varepsilon_{i,j\pm\frac{1}{2}} = \frac{1}{2}\left(\varepsilon_{ij} + \varepsilon_{i,j\pm1}\right). \tag{7.95}$$

We will take $\Delta x = \Delta y \equiv h$ and $N_x = N_y \equiv N$.

Define the residual by multiplying (7.94) by h^2:

$$F_{ij} \equiv \varepsilon_{i+\frac{1}{2},j}(u_{i+1,j} - u_{ij}) - \varepsilon_{i-\frac{1}{2},j}(u_{ij} - u_{i-1,j})$$
$$+ \varepsilon_{i,j+\frac{1}{2}}(u_{i,j+1} - u_{ij}) - \varepsilon_{i,j-\frac{1}{2}}(u_{ij} - u_{i,j-1}) \tag{7.96}$$

To solve $F_{ij}(u) = 0$, we will use Newton's method. Note the similarities with the Newton technique for solving nonlinear ODE boundary value problems.

We make a guess for u_{ij} that satisfies the Dirichlet boundary conditions and then iteratively change u_{ij} on the interior until the norm of the residual

$$\left\|F_{ij}^{(n)}\right\| \leq \epsilon \left\|F_{ij}^{(0)}\right\| \tag{7.97}$$

with $\epsilon = 10^{-9}$, for example, where $n = 1, 2, \ldots$ labels the Newton iterations. Set

$$u_{ij}^{(n)} = u_{ij}^{(n-1)} + \Delta u_{ij}, \quad i, j = 1, 2, \ldots, N-1. \tag{7.98}$$

For the initial guess, $u_{ij}^{(0)}$ may be taken to be zero on the interior.

For the linear algebra, we combine i and j into a single index $\alpha = 1, 2, \ldots, (N-1)^2$:

$$\alpha = (j-1)(N-1) + i, \quad i, j = 1, 2, \ldots, N-1. \tag{7.99}$$

Newton's equation for Δu_α is derived in the following way:

$$0 = F_\alpha\left(u^{(n)} + \Delta u\right) \approx F_\alpha\left(u^{(n)}\right) + \sum_\beta \left(\frac{\partial F_\alpha}{\partial u_\beta}\right)^{(n)} \Delta u_\beta \tag{7.100}$$

or

$$J^{(n)} \Delta u = -F\left(u^{(n)}\right) \tag{7.101}$$

7.8 Nonlinear Laplace Equation: Newton's Method

where the Jacobian $J_{\alpha\beta} = \partial F_\alpha / \partial u_\beta$. We solve (7.101) for Δu using a sparse matrix direct solve or GMRES,[7] and update $u \leftarrow u + \Delta u$.

Since the permittivity $\varepsilon_{ij} = \varepsilon(u_{ij})$ is a local (pointwise) function of the potential (see (7.95)),

$$\frac{\partial \varepsilon_{i \pm \frac{1}{2}, j}}{\partial u_{kl}} = \frac{1}{2} \left(\varepsilon'_{ij} \delta_{ik} \delta_{jl} + \varepsilon'_{i \pm 1, j} \delta_{i \pm 1, k} \delta_{jl} \right) \tag{7.102}$$

$$\frac{\partial \varepsilon_{i, j \pm \frac{1}{2}}}{\partial u_{kl}} = \frac{1}{2} \left(\varepsilon'_{ij} \delta_{ik} \delta_{jl} + \varepsilon'_{i, j \pm 1} \delta_{ik} \delta_{j \pm 1, l} \right) \tag{7.103}$$

where

$$\varepsilon'_{ij} = \frac{d\varepsilon}{du}(u_{ij}). \tag{7.104}$$

Thus, the explicit form of the Jacobian is

$$J_{ij,kl} = \frac{\partial F_{ij}}{\partial u_{kl}} = \delta_{ik} \delta_{jl}$$

$$\times \left(\frac{1}{2} \varepsilon'_{ij} (u_{i+1,j} + u_{i-1,j} + u_{i,j+1} + u_{i,j-1} - 4u_{ij}) \right.$$

$$\left. - \varepsilon_{i+\frac{1}{2},j} - \varepsilon_{i-\frac{1}{2},j} - \varepsilon_{i,j+\frac{1}{2}} - \varepsilon_{i,j-\frac{1}{2}} \right)$$

$$+ \delta_{i+1,k} \delta_{jl} \left(\frac{1}{2} \varepsilon'_{i+1,j} (u_{i+1,j} - u_{ij}) + \varepsilon_{i+\frac{1}{2},j} \right)$$

$$+ \delta_{i-1,k} \delta_{jl} \left(-\frac{1}{2} \varepsilon'_{i-1,j} (u_{ij} - u_{i-1,j}) + \varepsilon_{i-\frac{1}{2},j} \right)$$

$$+ \delta_{ik} \delta_{j+1,l} \left(\frac{1}{2} \varepsilon'_{i,j+1} (u_{i,j+1} - u_{ij}) + \varepsilon_{i,j+\frac{1}{2}} \right)$$

$$+ \delta_{ik} \delta_{j-1,l} \left(-\frac{1}{2} \varepsilon'_{i,j-1} (u_{ij} - u_{i,j-1}) + \varepsilon_{i,j-\frac{1}{2}} \right) \tag{7.105}$$

and then we convert $J_{ij,kl}$ to $J_{\alpha\beta}$.

As a realistic example in **nonlin_laplace.m**, we will take $\varepsilon(u) = 1 + au^2, a > 0$. Then $\varepsilon'(u) = 2au$.

---

[7] Since J is asymmetric, PCG is not applicable and GMRES should be used.

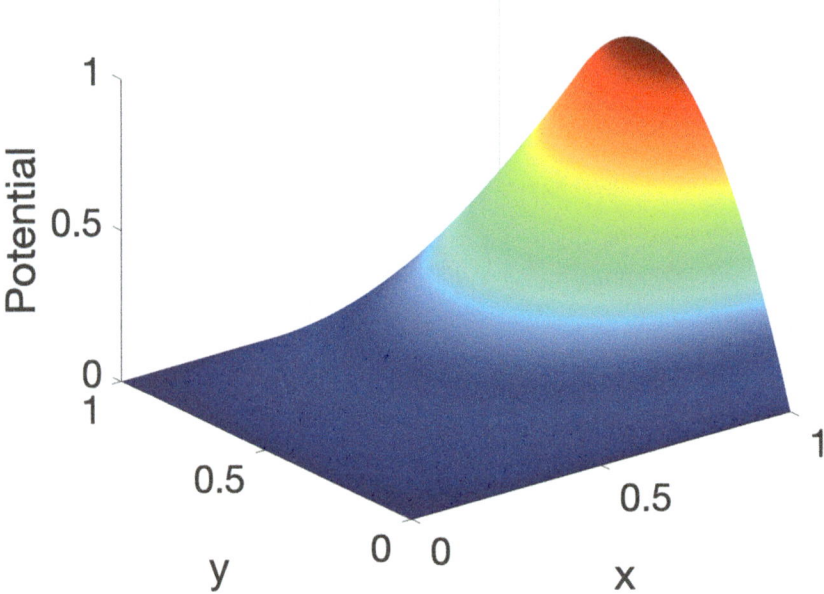

Fig. 7.7 Numerical solution to the nonlinear Laplace problem with $\varepsilon(u) = 1 + 2.5u^2$ on a $200\Delta x \times 200\Delta y$ grid with homogeneous Dirichlet boundary conditions except $u(x = 1, y) = 4y(1 - y)$. Compare with Fig. 7.1

7.8.2 Simulations of the Nonlinear Laplace Equation

A computation using **nonlin_laplace.m** for the nonlinear Laplace equation with homogeneous Dirichlet boundary conditions except $u(x = 1, y) = 4y(1 - y)$ is shown in Fig. 7.7. Note how the nonlinear solution, although largely controlled by the boundary conditions, rises along the midline $y = 0.5$ more linearly in x than the linear solution in Fig. 7.1, due to a higher permittivity $\varepsilon(u) = 1 + 2.5u^2 > 1$ away from the homogeneous Dirichlet boundaries, which reduces the magnitude of the electric field $\mathbf{E} = -\nabla u$.

The numerical methods in **nonlin_laplace.m** can be easily modified to solve the (nonlinear) Poisson-Boltzmann equation

$$\nabla^2 u = \sinh(u) \tag{7.106}$$

which describes the electrostatic potential energy in a neutral bath of ions (see Problem 7.11 for a 2D project). The nondimensional form is derived from the original Poisson-Boltzmann equation (see Sect. 9.3)

$$\nabla \cdot (\varepsilon \nabla \phi) = -\rho = -en_b \left(\exp\left\{ -\frac{e\phi}{k_B T} \right\} - \exp\left\{ \frac{e\phi}{k_B T} \right\} \right) \tag{7.107}$$

where $e > 0$ is the electronic charge, T is the ambient temperature, k_B is Boltzmann's constant, and where for simplicity we are considering one positive and one negative ion species with ionic charges $q = \pm e$ and equal ambient bath densities n_b. Note that the ion densities in the bath near thermal equilibrium are governed by the Boltzmann law

$$n = n_b \exp\left\{-\frac{q\phi}{k_B T}\right\}. \tag{7.108}$$

To nondimensionalize (7.107) and obtain (7.106), set the dimensionless potential energy $u = e\phi/(k_B T)$ and divide lengths by the Debye length for ions in the ambient bath

$$l_D = \sqrt{\frac{\varepsilon k_B T}{2e^2 n_b}} \approx 1 \text{ nm}. \tag{7.109}$$

Problems

7.1 For **nonlin_laplace.m**, an exact solution is not available. Estimate the order of accuracy for the sparse direct solve plus Newton method for the nonlinear Laplace equation by using a fine grid solution u_{fine} to stand in for the exact solution. Verify that our implementation achieves second-order accuracy in $\Delta x = \Delta y \equiv h$ by comparing computed solutions on spatial grids on $N\Delta x \times N\Delta y$ grids with $N = 100$, 200, and 400 with a fine grid solution with $N = 800$.

7.2 Show **laplace3.m** (SOR) converges under mesh refinement by comparing and contrasting simulations on $N\Delta x \times N\Delta y$ grids with $\Delta x = \Delta y = h$ and $N = 50$, 100, and 200.

7.3 Show **nonlin_laplace.m** (Newton's method) converges under mesh refinement by comparing and contrasting simulations on $N\Delta x \times N\Delta y$ grids with $\Delta x = \Delta y \equiv h$ and $N = 50$, 100, and 200.

7.4 Show the Jacobi spectral radius $\rho_J = \cos(\pi h)$ for Laplace's equation on the unit square with second-order accurate central differences. Write up only the 1D version. *Hint:* In 1D, set $A = \text{tridiag}[-1\ 2\ -1]$. The iteration matrix $B = \frac{1}{2}\text{tridiag}[1\ 0\ 1]$. Then show $\left(Bv^{(j)}\right)_k = \cos(j\pi h) v_k^{(j)}$, $j, k = 1, 2, \ldots, n$, where here $h = 1/(n+1)$ and the 1D eigenvectors are

$$v^{(j)} = [\sin(j\pi h), \sin(2j\pi h), \cdots, \sin(nj\pi h)]$$

or $v_k^{(j)} = \sin(jk\pi h)$. The Jacobi spectral radius $\rho_J = \max_j |\cos(j\pi h)| = \cos(\pi h)$. Be sure to verify that your calculation works for the boundaries $k = 1$ and $k = n$.

7.5

(a) For SOR/SUR, show $\det\{B\} = (1-\omega)^n, 0 < \omega < 2$. *Hint: B is the product of triangular matrices* $(A = L + D + U)$:

$$B = (D + \omega L)^{-1}((1-\omega)D - \omega U).$$

(b) Derive the equation for the SOR ω_{opt}

$$\omega_{opt} = \frac{2\left(1 - \sqrt{1-\mu^2}\right)}{\mu^2}, \quad \lambda = \omega_{opt} - 1$$

assuming Young's formula applied to the spectral radii:

$$(\lambda + \omega - 1)^2 = \lambda \omega^2 \mu^2.$$

Here μ is the spectral radius for the Jacobi iteration method, and λ is the spectral radius for the SOR iteration method. *Hint:* Set $\lambda = \omega - 1$ in Young's formula, find $\omega_\pm$ using the quadratic formula, and then minimize λ by choosing the correct sign for ω_{opt}.

(c) Show for Laplace's equation on the unit square; the SOR spectral radius is

$$\lambda = \frac{1 - \sin \pi h}{1 + \sin \pi h}.$$

7.6

(a) Discretize the 3D Laplace equation

$$\nabla^2 u(x, y, z) = u_{xx} + u_{yy} + u_{zz} = 0$$

with second-order accurate central differences and $\Delta x = \Delta y = \Delta z \equiv h$.

(b) Then write down a formula for u_{ijk} in terms of its six nearest neighbors.

(c) Write down the Jacobi iteration formula for $u_{ijk}^{(n+1)}$ at iteration $n+1$ in terms of the six nearest neighbor values $u_{NN}^{(n)}$ at iteration n.

7.7

(a) Discretize the 2D Poisson equation

$$\nabla^2 u(x, y) = u_{xx} + u_{yy} = -\rho(x, y)$$

where $\rho(x, y)$ is the given charge density, with second-order accurate central differences and $\Delta x = \Delta y \equiv h$.

(b) Express u_{ij} in terms of its four nearest neighbors and ρ_{ij}.

Problems 119

(c) Write down the Jacobi iteration formula for $u_{ij}^{(k+1)}$ at iteration $k+1$ in terms of the four nearest neighbor values $u_{NN}^{(k)}$ at iteration k, plus ρ.

(d) Define the residual $r_{ij} = h^2\left((\nabla^2 u)_{ij} + \rho_{ij}\right)$. Then write the Jacobi method in terms of the residual $r_{ij}^{(k)}$.

Finally write down the (e) Gauss-Seidel and (f) SOR iteration formulas for $u_{ij}^{(k+1)}$. Use the notation $\tilde{r}$ to denote the residual with current values $\tilde{u}$ of u (either updated or not) on the right-hand side.

7.8 For conjugate gradients, prove the formulas (7.79) and (7.82) for the α_j and β_j. *Hint:* Use induction.

7.9 Justify each of the steps (7.87)–(7.90) in proving the GMRES algorithm.

7.10 Experiment with the GMRES solver in **laplace6.m** by trying different choices of parameters in

    gmres(A,b,RESTART,EPSILON,MAXIT,M1,M2)

like

    gmres(A,b,50,10^-6,1000,L,L')

with

    L = ichol(A,struct('michol','on'));

etc.

7.11 *Project:* Modify **nonlin_laplace.m** to solve the 2D Poisson-Boltzmann equation (7.106) on the rectangle $[1, 5] \times [-1, 1]$ with boundary conditions

$$u(x = 1, y) = (1 - y^2)^2, \quad u_y(x, y = \pm 1) = 0, \quad u(x = 5, y) = 0.$$

Chapter 8
Numerical Methods for Hyperbolic PDEs

Numerical methods for *hyperbolic* PDEs like the wave equations, Maxwell's equations of electromagnetism, the inviscid Burgers equation, gas dynamics, and magnetohydrodynamics are completely different from numerical methods for parabolic PDEs. In hyperbolic PDEs, information propagates in the form of waves with finite velocity. Numerical methods for hyperbolic PDEs will have to take this wave propagation into account: for example, by upwinding spatial derivatives.

Our main analytical tools for hyperbolic PDEs are the classification of waves (shock waves, contact waves, and rarefaction waves in gas dynamics) and of Riemann invariants, which are conserved along characteristic curves in the space-time plane, and exact solutions to Riemann problems, i.e., solutions to 1D initial value problems with constant data to the left and to the right of the origin. Additionally, for nonlinear hyperbolic PDEs—which can develop discontinuous solutions even from smooth initial conditions—we utilize the Rankine-Hugoniot jump conditions at discontinuities.

Mathematically appropriate boundary conditions for hyperbolic PDEs are Cauchy, which are based on how information propagates with finite velocity along characteristics in hyperbolic problems.

The simplest hyperbolic PDE is the first-order wave (or linear advection) equation

$$u_t + c u_x = 0, \quad u(x, t = 0) = u_0(x), \quad c > 0 \tag{8.1}$$

where $u(x, t)$ is the amplitude of the wave. The initial wave form $u_0(x)$ simply propagates unchanged to the right with velocity c:

$$u(x, t) = u_0(x - ct). \tag{8.2}$$

Thus, when discretizing the first-order wave equation, we can use the *upwind* method that looks upstream (to the left) to discretize u_x:

$$\frac{u_i^{n+1} - u_i^n}{\Delta t} = -c\frac{u_i^n - u_{i-1}^n}{\Delta x} \tag{8.3}$$

$$u_i^{n+1} = u_i^n - \frac{c\Delta t}{\Delta x}(u_i^n - u_{i-1}^n) \tag{8.4}$$

where the exact solution $u(x_i, t_n)$ is approximated by u_i^n.

Stability and accuracy of hyperbolic methods can be analyzed by means of the *modified PDE*, where we take the numerical method and then Taylor expand to see the leading effect of the numerical method on the PDE. For the upwind method, we obtain the modified PDE

$$u + \Delta t\, u_t + \frac{\Delta t^2}{2}u_{tt} + \cdots = u - c\Delta t\left(u_x - \frac{\Delta x}{2}u_{xx} + \cdots\right) \tag{8.5}$$

or

$$u_t + cu_x = \frac{c\Delta x}{2}u_{xx} - \frac{\Delta t}{2}u_{tt} + \cdots \approx \frac{c}{2}(\Delta x - c\Delta t)u_{xx} \equiv D_n u_{xx} \tag{8.6}$$

where D_n is the coefficient of numerical diffusion, making use of the fact that $u_t \approx -cu_x$ to leading order. The form of (8.6) shows the upwind method is first-order accurate and has a diffusive effect on numerical solutions. To be stable, Δt must be $\leq \Delta x/c$, since D_n must be ≥ 0.

Table 8.1 lists the numerical methods we will discuss in this chapter.

First-order hyperbolic methods (like upwind and Lax-Friedrichs) are diffusive, since they spread out propagating waves through numerical diffusion. Second-order hyperbolic methods (like Lax-Wendroff) are dispersive, since they introduce dispersive oscillations in propagating waves. Upwind methods are preferred, especially modern high-order upwind methods that minimize the amount of numerical diffusion and dispersion.

All of the standard hyperbolic methods are explicit, since a small timestep is necessary in any case to resolve wave interactions; also implicit methods for hyperbolic PDEs are typically far too diffusive or dispersive.

As a general rule, explicit hyperbolic methods must obey the CFL (Courant-Friedrichs-Lewy) timestep restriction for stability

Table 8.1 Numerical methods for hyperbolic PDEs. All the methods are stable for $\Delta t \leq \Delta x / \max\{|\lambda|\}$, where $\max\{|\lambda|\}$ is the maximum characteristic speed in magnitude

Method	Type	Order	Effect
Upwind	Explicit	First	Diffusive
Lax-Friedrichs	Explicit	First	Diffusive
Lax-Wendroff	Explicit	Second	Dispersive
Godunov	Explicit	First	Diffusive
WENO3	Explicit	Third	Diffusive

8 Numerical Methods for Hyperbolic PDEs

$$\Delta t \leq \frac{\Delta x}{\max\{|\lambda|\}} \tag{8.7}$$

where $\max\{|\lambda|\}$ is the maximum characteristic speed in magnitude on the grid at the beginning of the timestep. In certain cases, there may be a fraction like $\frac{1}{2}$ multiplying $\Delta x / \max\{|\lambda|\}$. For hyperbolic PDEs, Δt will always be adjusted dynamically based on the CFL condition.

The modern approach is to solve systems of first-order (in time) PDEs. Thus, to solve the wave equation

$$w_{tt} - c^2 w_{xx} = 0 \tag{8.8}$$

we transform to two coupled first-order (in time) PDEs by setting $u = cw_x$ and $v = w_t$:

$$u_t = cv_x, \quad v_t = cu_x \tag{8.9}$$

and then solve each first-order (in time) PDE with our standard methods.

We are mainly concerned with hyperbolic PDEs in the form of conservation laws

$$w_t + f(w)_x = 0 \tag{8.10}$$

where w is the set of conserved quantities with fluxes $f(w)$. In general, w and f will have m components; in 1D gas dynamics, for example, $m = 3$.

Our numerical methods for hyperbolic conservation laws should conserve (in exact arithmetic) quantities that are conserved for the original continuous PDEs, so we always use *conservative* numerical methods. Conservative methods for (8.10) can be written in the form

$$w_i^{n+1} = w_i^n - \frac{\Delta t}{\Delta x} \left(F_{i+\frac{1}{2}} - F_{i-\frac{1}{2}} \right) \tag{8.11}$$

where the numerical flux function $F = F(w^n, w^{n+1}, \Delta t)$ describes the numerical method (see Sect. 6.3). For explicit methods, $F = F(w^n, \Delta t) \equiv F^n$.

For linear hyperbolic PDEs, we typically get the exact solution (projected onto the spatial grid) for the classical schemes like upwind, Lax-Friedrichs, and Lax-Wendroff if the Courant number r is set to 1 in

$$\Delta t = r \frac{\Delta x}{\max\{|\lambda|\}}, \quad 0 < r \leq 1. \tag{8.12}$$

For nonlinear hyperbolic PDEs, $r = 1$ is often computationally unstable (although theoretically stable in the $\Delta x \to 0$ limit), and usually we take $0.1 \leq r \leq 0.9$. When different Courant numbers r give stable results for the nonlinear problem, the solutions generally are the same graphically to within a line width. For linear hyperbolic PDEs, $r < 1$ leads to excessive diffusion or dispersion of waves.

In numerical codes, the Courant number r is usually called the *CFL factor*, and the timestep is given by

$$\Delta t = CFL \frac{\Delta x}{\max\{|\lambda|\}}, \quad 0 < CFL \leq 1. \tag{8.13}$$

The methods of choice for nonlinear hyperbolic problems like gas dynamics are WENO [33] or higher-order Godunov methods like PPM [9] and CLAWPACK [25]. Here we focus on WENO, but Godunov's method is also presented and discussed.

The Finite-Difference Time-Domain (FDTD) method (see [37] for example) should be used for the linear Maxwell equations.

For another general introduction to solving hyperbolic PDEs, see LeVeque's *Finite Difference Methods for Ordinary and Partial Differential Equations* [26]. For more advanced treatments of hyperbolic conservation laws, see Chorin and Marsden's *A Mathematical Introduction to Fluid Mechanics* [8] and Smoller's *Shock Waves and Reaction-Diffusion Equations* [34] for theory and LeVeque's *Finite Volume Methods for Hyperbolic Problems* [25] for numerical methods. For a comprehensive treatment of linear and nonlinear waves, see Whitham's *Linear and Nonlinear Waves* [40].

8.1 First-Order Wave (Linear Advection) Equations

There are two *first-order wave (or linear advection) equations*, one propagating to the right (upper sign) and one propagating to the left (lower sign):

$$\frac{\partial u}{\partial t} \pm c \frac{\partial u}{\partial x} = 0, \quad u(x, t = 0) = u_0(x), \quad c > 0 \tag{8.14}$$

where $u(x, t)$ is the amplitude of the wave.

We will focus on the first-order wave equation propagating to the right:

$$\frac{\partial u}{\partial t} + c \frac{\partial u}{\partial x} = 0, \quad u(x, t = 0) = u_0(x), \quad c > 0. \tag{8.15}$$

Results for the first-order wave equation propagating to the left can be obtained by the symmetry $c \to -c$.

The solution to the initial value problem (8.15)

$$u(x, t) = u_0(x - ct) \tag{8.16}$$

8.1 First-Order Wave (Linear Advection) Equations

propagates the initial data u_0 to the right with velocity c. To verify this solution, note that $u(x, t = 0) = u_0(x)$ and that

$$u_t - cu_x = -cu'_0 + cu'_0 = 0. \tag{8.17}$$

Note that going backward in time $t \to -t$ simply propagates the initial data to the left with velocity $-c$ in (8.15).

The *Riemann invariant* u is constant along characteristics λ (see Fig. 8.1) with $x = x_0 + ct$ and $u(x, t) = u_0(x_0)$ since

$$\left.\frac{du}{dt}\right|_\lambda = \frac{\partial u}{\partial t} + \frac{dx}{dt}\frac{\partial u}{\partial x} = \frac{\partial u}{\partial t} + c\frac{\partial u}{\partial x} = 0. \tag{8.18}$$

Without the boundaries in Fig. 8.1, the solution u is determined at both *dots* by the initial data at the corresponding feet of the characteristics. *With* the boundaries, the solution u is determined at the right *dot* by the initial datum at the foot of the characteristic, but the solution at the left *dot* must be determined by a boundary condition at the left. So with boundaries, the appropriate Cauchy boundary condition is to specify u along the left boundary.

The simplest computational boundary conditions for hyperbolic PDEs are periodic (Sect. 8.20.1) or through-flow (Sect. 8.20.2), both of which are implemented in our first-order **wave1.m** and second-order **wave2.m** wave equation programs.

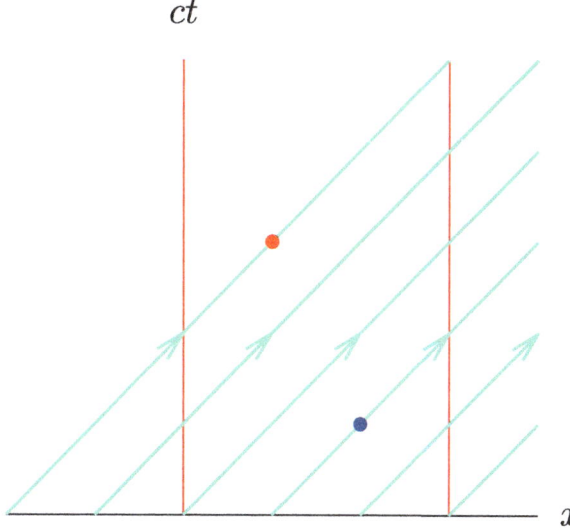

Fig. 8.1 Characteristics for the first-order wave equation $u_t + cu_x = 0$

8.2 Second-Order Wave Equation

The (second-order) *wave equation* is

$$\frac{\partial^2 w}{\partial t^2} - c^2 \frac{\partial^2 w}{\partial x^2} = 0, \quad w(x, t = 0) = w_0(x), \quad w_t(x, t = 0) = v_0(x) \quad (8.19)$$

where $w(x, t)$ is the amplitude (or pressure difference from ambient, etc.) of the wave. The solution propagates the initial data to the left and right with velocities $\pm c$:

$$w(x, t) = f(x - ct) + g(x + ct). \quad (8.20)$$

Note that

$$\begin{aligned} w_{tt} - c^2 w_{xx} &= c^2 f''(x - ct) + c^2 g''(x + ct)) \\ &\quad - c^2 (f''(x - ct) + g''(x + ct)) = 0. \end{aligned} \quad (8.21)$$

Again going backward in time $t \to -t$ simply propagates the initial data oppositely to the left and right with velocities $\mp c$.

The Riemann invariant f is constant along characteristics with $dx/dt = c$ (*light lines* in Fig. 8.2), and the Riemann invariant g is constant along characteristics with $dx/dt = -c$ (*dark lines*).

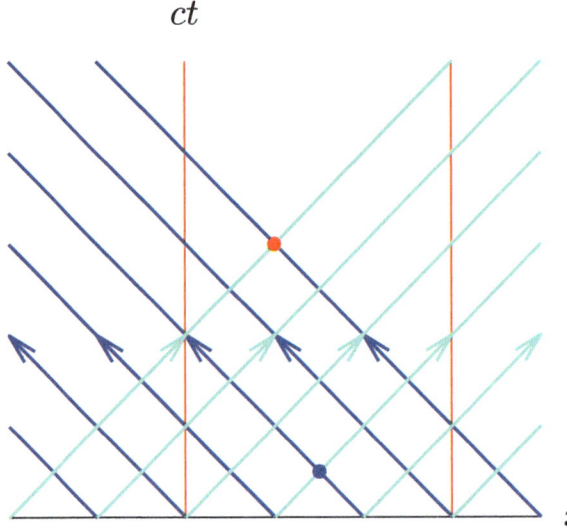

Fig. 8.2 Characteristics for the second-order wave equation $u_{tt} - c^2 u_{xx} = 0$

8.3 Transformation to Characteristic Variables

Without the boundaries in Fig. 8.2, the solution $u = f + g$ is determined at both *dots* by the initial data at the corresponding feet of the f and g characteristics. *With* the boundaries, the solution u is determined at the *right dot* by the initial data at the feet of the f and g characteristics, but the solution at the *left dot* must be determined by one boundary condition at the left (for f) and one boundary condition at the right (for g). So with boundaries, the appropriate Cauchy boundary conditions are to specify u along the left and right boundaries.

The D'Alembert solution to the wave equation expresses the Riemann invariants f and g in terms of the initial conditions:

$$w(x,t) = \frac{1}{2}w_0(x - ct) + \frac{1}{2}w_0(x + ct) + \frac{1}{2c}\int_{x-ct}^{x+ct} v_0(s)\,ds. \tag{8.22}$$

To solve the wave equation numerically, we convert it to a first-order system ($u = cw_x$, $v = w_t$):

$$u_t = cv_x \tag{8.23}$$

$$v_t = cu_x \tag{8.24}$$

with $u(x, t = 0) = cw_0'(x)$, $v(x, t = 0) = v_0(x)$. This splitting is explored in **wave2.m**.

Alternatively, for upwinding, we can now set $U = u - v$ and $V = u + v$ to obtain

$$U_t + cU_x = 0 \tag{8.25}$$

$$V_t - cV_x = 0 \tag{8.26}$$

with $U(x, t = 0) = cw_0'(x) - v_0(x)$, $V(x, t - 0) = cw_0'(x) + v_0(x)$. Equation (8.25) propagates waves to the right, and (8.26) propagates waves to the left. This transformation is an example of using characteristic variables (see the next Sect. 8.3).

8.3 Transformation to Characteristic Variables

First consider the linear system of hyperbolic conservation laws

$$w_t + Aw_x = 0 \tag{8.27}$$

where w has m components and A is an $m \times m$ constant matrix. The system (8.27) is hyperbolic if A is diagonalizable with real eigenvalues:

$$A = R\Lambda R^{-1} \tag{8.28}$$

where $\Lambda = \text{diag}\{\lambda_1, \lambda_2, \ldots, \lambda_m\}$ is the matrix of eigenvalues of A and $R = [r_1 \; r_2 \; \ldots r_m]$ is the matrix of right eigenvectors of A. The system (8.27) is strictly hyperbolic if the eigenvalues are real and distinct. Define the *characteristic variables* $u = R^{-1}w$, and multiply (8.27) by R^{-1} (R^{-1} is the left eigenvector matrix of A) to obtain

$$u_t + \Lambda u_x = 0 \tag{8.29}$$

since R^{-1} is constant. This linear system of equations can be written as

$$u_{\alpha t} + \lambda_\alpha u_{\alpha x} = 0, \quad \alpha = 1, 2, \ldots, m \tag{8.30}$$

which has the exact solution

$$u_\alpha(x, t) = u_\alpha(x - \lambda_\alpha t, 0) \tag{8.31}$$

where $u(x, 0) = R^{-1}w_0(x)$. Finally $w(x, t) = Ru(x, t)$.

The straight lines $x = x_0 + \lambda_\alpha t$ with $dx/dt = \lambda_\alpha$ are the characteristics of the αth family or α-characteristics. More generally, the characteristics will be curves in the x-t plane.

For nonlinear systems of hyperbolic conservation laws $w_t + f(w)_x = 0$ like gas dynamics, we will transform locally to characteristic variables before applying higher-order upwind methods (see Sect. 8.23), using the local left and right eigenvectors of the Jacobian $\partial f/\partial w$ at grid points $i \pm \frac{1}{2}$. In this way, more difficult problems involving stronger waves can be computed.

8.4 First-Order Upwind Method

The first successful numerical method for hyperbolic PDEs was the *upwind* method (Courant et al.) [11]. For $u_t + cu_x = 0$, the upwind method is

$$\frac{u_i^{n+1} - u_i^n}{\Delta t} = -c\frac{u_i^n - u_{i-1}^n}{\Delta x} \tag{8.32}$$

or in standard form (unfolding the time derivative)

$$u_i^{n+1} = u_i^n - c\frac{\Delta t}{\Delta x}(u_i^n - u_{i-1}^n). \tag{8.33}$$

It is first-order accurate and *CFL stable*, i.e., stable for $\Delta t \leq \Delta x/c$ (CFL condition). It captures the *physics* of wave propagation by differencing upstream at the old time level.

Consistency: The LTE ($= \Delta t \, \tau$) is given by

8.5 Modified Equation and Stability

$$u(x, t + \Delta t) = u(x, t) - \frac{c\Delta t}{h}(u(x, t) - u(x - h, t)) + \Delta t\, \tau. \tag{8.34}$$

Taylor expanding, we get

$$u + \Delta t\, u_t + \frac{\Delta t^2}{2} u_{tt} + \cdots = u - c\Delta t \left(u_x - \frac{h}{2} u_{xx} + \cdots \right) + \Delta t\, \tau \tag{8.35}$$

$$\tau = \frac{\Delta t}{2} u_{tt} - \frac{ch}{2} u_{xx} + \cdots \tag{8.36}$$

For $r = 1$, the upwind method produces the exact solution projected onto the spatial grid:

$$u_i^{n+1} = u_i^n - (u_i^n - u_{i-1}^n) = u_{i-1}^n = u(x_i - c\Delta t, t_n). \tag{8.37}$$

Similar remarks apply to Eqs. (8.25) and (8.26) to show upwinding produces the exact solution projected onto the grid for $r = 1$ for the second-order wave equation.

Problem 8.8 confirms that the Lax-Friedrichs and Lax-Wendroff methods also produce the exact solution projected onto the grid for $r = 1$ for the first-order and second-order wave equations.

The first-order upwind method for $u_t + cu_x = 0$, with $c = 1$, is implemented in our MATLAB code **wave1.m**.

8.5 Modified Equation and Stability

To analyze stability, we will derive the *modified equation* for the upwind method. In the Taylor series expansion above with the LTE on the right-hand side, we delete the LTE and then do *not* use the wave equation:

$$u + \Delta t\, u_t + \frac{\Delta t^2}{2} u_{tt} + \cdots = u - c\Delta t \left(u_x - \frac{h}{2} u_{xx} + \cdots \right) \tag{8.38}$$

or

$$u_t + cu_x = \frac{ch}{2} u_{xx} - \frac{\Delta t}{2} u_{tt} + \cdots \approx \frac{c}{2}(h - c\Delta t) u_{xx} = \frac{ch}{2}(1 - r) u_{xx} \equiv D_n u_{xx} \tag{8.39}$$

where $r = c\Delta t/h$ is the Courant number, and making use of the fact that $u_t \approx -cu_x$ to leading order. This modified equation shows the leading order effects of the numerical method on the original PDE. The modified equation is stable for $\Delta t \leq h/c$, since in that case the numerical diffusion coefficient $D_n \geq 0$. (This derivation is not strictly speaking a proof of stability, but it does strongly suggest upwind is stable for the wave equation.) Note that upwind is diffusive since the leading order

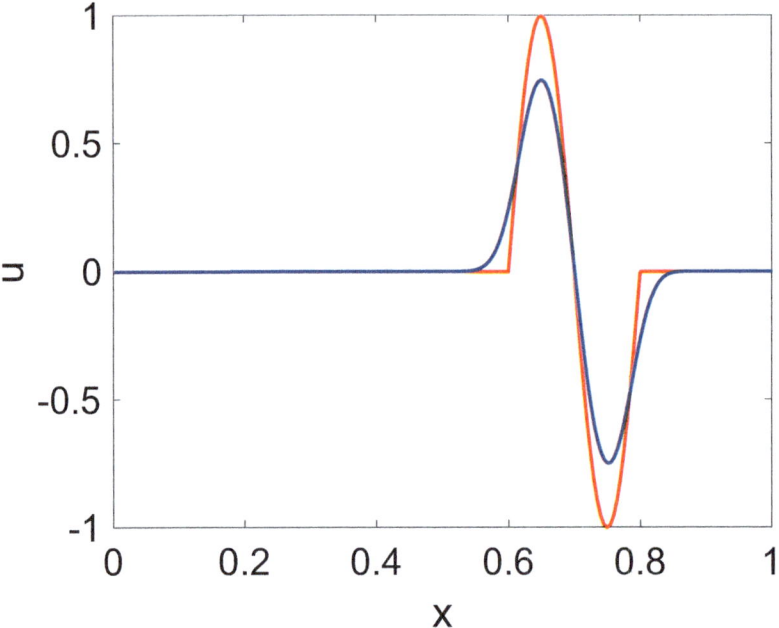

Fig. 8.3 Upwind solutions propagating to the right at speed $c = 1$ to the first-order wave equation computed using **wave1.m** with $200\Delta x$ at $t = 0.6$: exact ($r = 1$) solution projected onto the grid and diffusive solution ($r = 0.8$)

effect of the method on the wave equation is to introduce the diffusive u_{xx} term (see Fig. 8.3). Higher-order spatial derivative terms on the right-hand side of (8.39) will go to zero faster than the diffusive term as $\Delta t \to 0$ and $\Delta x \to 0$, since the jth spatial derivative of u will be multiplied by Δx^{j-1}, where $j = 2, 3, \ldots$, using $\Delta t \sim \Delta x$ for hyperbolic methods.

Requiring $\Delta t \leq h/c$ for stability is called the *CFL condition*. It says *the domain of dependence of the PDE must be included in the domain of dependence of the difference method*. The CFL condition is necessary but not sufficient for stability (see the FTCS scheme below).

The downwind method (which violates the CFL condition and is only used by mistake)

$$u_i^{n+1} = u_i^n - c\frac{\Delta t}{\Delta x}(u_{i+1}^n - u_i^n) \tag{8.40}$$

is unconditionally unstable, since now for the modified equation we have

$$u_t + cu_x = -\frac{ch}{2}u_{xx} - \frac{\Delta t}{2}u_{tt} + \cdots \approx -\frac{c}{2}(h + c\Delta t)u_{xx}$$

$$= -\frac{ch}{2}(1+r)u_{xx} \equiv D_n u_{xx} \tag{8.41}$$

and $D_n < 0$, making the modified equation equivalent to the (always unstable) backward diffusion equation.

Similarly, the central difference FTCS (forward time central space) scheme

$$u_i^{n+1} = u_i^n - c\frac{\Delta t}{2\Delta x}(u_{i+1}^n - u_{i-1}^n) \tag{8.42}$$

is unconditionally unstable. For the modified equation, we have

$$u + \Delta t\, u_t + \frac{\Delta t^2}{2} u_{tt} + \cdots = u - c\Delta t \left(u_x + \frac{h^2}{6} u_{xxx} + \cdots \right) \tag{8.43}$$

or to leading order

$$u_t + cu_x = -\frac{c^2 \Delta t}{2} u_{xx} \equiv D_n u_{xx} \tag{8.44}$$

and $D_n < 0$.

8.6 Von Neumann Stability Analysis

Upwind Method for $u_t + cu_x = 0$. Set u_j^n equal to a single Fourier mode $u_j^n = e^{ikx_j}$, and derive the growth factor $u_j^{n+1} = G(k)u_j^n$ for this mode:

$$u_j^{n+1} = u_j^n - r(u_j^n - u_{j-1}^n), \quad r = \frac{c\Delta t}{h} \tag{8.45}$$

$$u_j^{n+1} = \left(1 - r\left(1 - e^{-ikh}\right)\right) e^{ikx_j} = G(k)u_j^n \tag{8.46}$$

$$G(k) = 1 - r + re^{-ikh} \tag{8.47}$$

$$|G(k)|^2 = (1-r)^2 + r^2 + r(1-r)\left(e^{-ikh} + e^{ikh}\right)$$
$$= 2r^2 - 2r + 1 + 2r(1-r)\cos(kh) \leq 1 \tag{8.48}$$

iff

$$(r-1)(1 - \cos(kh)) \leq 0 \quad \text{iff} \quad r \leq 1. \tag{8.49}$$

Thus, upwind is stable if $\Delta t \leq \Delta x/c$.

FTCS for $u_t + cu_x = 0$ satisfies the CFL condition but is unconditionally unstable:

$$u_j^{n+1} = u_j^n - \frac{r}{2}(u_{j+1}^n - u_{j-1}^n) \tag{8.50}$$

$$u_j^{n+1} = \left(1 - \frac{r}{2}\left(e^{ikh} - e^{-ikh}\right)\right)e^{ikx_j} = G(k)u_j^n \tag{8.51}$$

$$G(k) = 1 - ir\sin(kh) \tag{8.52}$$

$$|G(k)|^2 = 1 + r^2\sin^2(kh) > 1 \tag{8.53}$$

whenever $\sin(kh) \neq 0$.

8.7 Lax-Friedrichs Method

The *Lax-Friedrichs (LF)* method for $u_t + cu_x = 0$ redeems the central stencil approach by omitting u_i^n on the right-hand side in favor of the average $\frac{1}{2}(u_{i-1}^n + u_{i+1}^n)$:

$$u_i^{n+1} = \frac{1}{2}(u_{i-1}^n + u_{i+1}^n) - c\frac{\Delta t}{2\Delta x}(u_{i+1}^n - u_{i-1}^n). \tag{8.54}$$

LF is first-order accurate, stable for $\Delta t \leq \Delta x/c$, and conservative (see Sect. 8.8). Replacing the central point on the right-hand side with the average introduces enough numerical diffusion to stabilize the FTCS scheme. Note that Lax-Friedrichs is too diffusive (spreading out propagating waves too much), so modern state-of-the-art hyperbolic methods are all higher-order upwind methods—which minimize the amount of numerical diffusion and dispersion. However, Lax-Friedrichs, intermixed with a high-order upwind scheme, still plays an important role in positivity-preserving schemes for gas dynamics (see Sect. 8.23).

The modified equation for LF for $u_t + cu_x = 0$

$$u_t + cu_x = \frac{ch}{2r}(1 - r^2)u_{xx} \equiv D_n u_{xx}, \quad r = \frac{c\Delta t}{h} \tag{8.55}$$

shows it is diffusive (see Fig. 8.4). Note that $D_n \geq 0$ when $r \leq 1$.

The Lax-Friedrichs method is implemented in our MATLAB code **wave2.m** for the second-order wave equation, written as a two-equation first-order system (8.23) and (8.24).

For the nonlinear hyperbolic conservation law $w_t + f(w)_x = 0$, the LF scheme is

8.8 Hyperbolic Conservation Laws

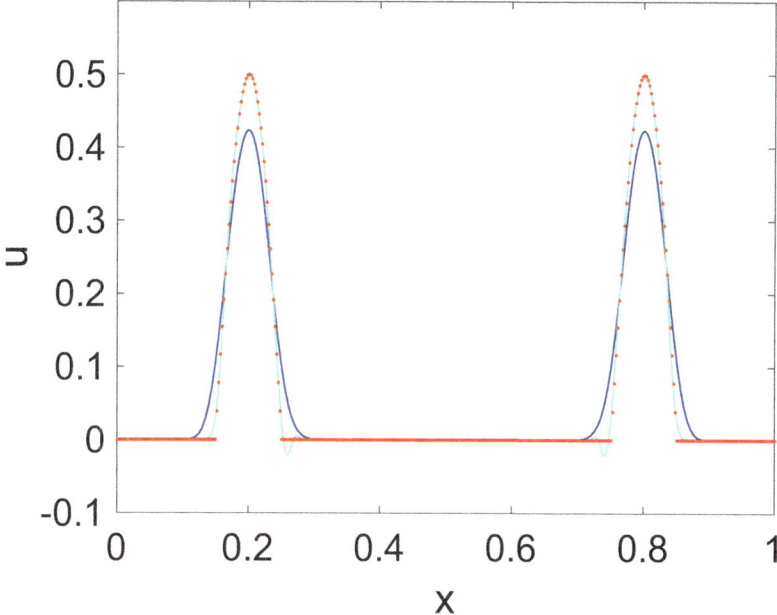

Fig. 8.4 Diffusive Lax-Friedrichs (*lower curve*, $r = 0.8$) and dispersive Lax-Wendroff (*upper curve*, $r = 0.8$) solutions to the second-order wave equation computed using **wave2.m** with $400\Delta x$ at $t = 0.3$, compared with the Lax-Friedrichs exact solution (*dots*, $r = 1$) projected onto the grid. Here the left (right) wave is traveling to the left (right) with velocity $-c = -1$ ($+c = 1$), and the dispersive LW oscillations trail the waves

$$w_i^{n+1} = \frac{1}{2}(w_{i-1}^n + w_{i+1}^n) - \frac{\Delta t}{2\Delta x}(f(w_{i+1}^n) - f(w_{i-1}^n)). \tag{8.56}$$

There are four different versions of the Lax-Friedrichs method. The original plus three modifications will be discussed in Sect. 8.15 and numerical results compared for the inviscid Burgers equation.

8.8 Hyperbolic Conservation Laws

In 1D, a system of microscopic conservation laws is written as

$$w_t + f(w)_x = 0 = w_t + \left(\frac{\partial f}{\partial w}\right) w_x \tag{8.57}$$

where the conserved variables $w(x, t)$ and flux functions $f(w)$ have m components. The system of macroscopic conservation laws for the m conserved quantities $Q =$

$\int_a^b w\,dx$ is

$$\frac{dQ}{dt} = \int_a^b w_t\,dx = -\int_a^b f(w)_x\,dx = f(w(a)) - f(w(b)) = \text{inflow} - \text{outflow}. \tag{8.58}$$

The 1D system of conservation laws (8.57) is *hyperbolic* if the Jacobian matrix $\partial f/\partial w$ has real eigenvalues and is diagonalizable (with a complete set of linearly independent eigenvectors). Then solutions to the system of conservation laws can be viewed in terms of propagating waves. If the eigenvalues of $\partial f/\partial w$ are real and distinct, then the eigenvectors are guaranteed to form a complete linearly independent set, and the conservation law is called *strictly hyperbolic*.

For hyperbolic conservation laws, the Lax-Friedrichs, Lax-Wendroff, ENO, WENO, and Godunov (PPM, CLAWPACK, etc.) methods are conservative, while the original upwind method is not (original upwind is conservative for $u_t + (c(x,u)u)_x = 0$ only if c does not change sign). Two-step Lax-Wendroff is manifestly in conservation form. For Lax-Friedrichs, the numerical flux function for $w_t + f(w)_x = 0$ is

$$F^n_{i+\frac{1}{2}} = \frac{1}{2}(f(w^n_i) + f(w^n_{i+1})) - \frac{\Delta x}{2\Delta t}(w^n_{i+1} - w^n_i). \tag{8.59}$$

In 3D, a system of microscopic conservation laws is written as

$$w_t + \nabla \cdot \mathbf{f} = 0, \quad \mathbf{f} = (f, g, h) \tag{8.60}$$

where the conserved variables $w(x,t)$ and flux functions $f(w)$, $g(w)$, and $h(w)$ have m components. The system of macroscopic conservation laws for the m conserved quantities $Q = \int_V w\,dV$ is

$$\frac{dQ}{dt} = \int_V \frac{\partial w}{\partial t}\,dV = -\int_{\partial V} \mathbf{f} \cdot \mathbf{da} = \text{inflow} - \text{outflow}. \tag{8.61}$$

To classify the 3D system of conservation laws (8.60), it is best to calculate the eigenvalues of the symbol of the linearized PDE system (see Sect. 5.2 and Problem 9.4).

8.9 Lax-Wendroff Method

The *Lax-Wendroff (LW)* method is a second-order accurate central scheme but is too dispersive to be used to advantage in scientific computations, and higher-order upwind methods (like ENO/WENO, PPM, or CLAWPACK) are preferred.

Nonetheless LW is important historically and as a building block for constructing some modern methods.

The *Lax-Wendroff (LW)* method for $u_t + cu_x = 0$

$$u_i^{n+1} = u_i^n - c\frac{\Delta t}{2\Delta x}(u_{i+1}^n - u_{i-1}^n) + \frac{1}{2}c^2\frac{\Delta t^2}{\Delta x^2}(u_{i+1}^n - 2u_i^n + u_{i-1}^n) \qquad (8.62)$$

is stable for $\Delta t \leq \Delta x/c$ and conservative. It is derived by Taylor expanding

$$u(t + \Delta t) \approx u + \Delta t\, u_t + \frac{\Delta t^2}{2}u_{tt} = u - c\Delta t\, u_x + c^2\frac{\Delta t^2}{2}u_{xx} \qquad (8.63)$$

using $u_t = -cu_x$ and then replacing the x derivatives with the three-point central difference approximations. Since LW agrees with the Taylor series expansion through second order (and we used second-order accurate central derivatives in x), it is second-order accurate. The modified equation for LW is

$$u_t + cu_x = \frac{ch^2}{6}(r^2 - 1)u_{xxx} - \epsilon u_{xxxx} \qquad (8.64)$$

with $\epsilon = \frac{ch^3}{8}r(1 - r^2) > 0$ when $r < 1$, suggesting it is CFL stable. Note that LW is dispersive, since the leading order effect of the method on the wave equation is to introduce the dispersive u_{xxx} term (see Fig. 8.4).

8.10 Two-Step Lax-Wendroff Method

For nonlinear hyperbolic conservation laws $w_t + f(w)_x = 0$, the *two-step Lax-Wendroff* method should be used. (Two-step Lax-Wendroff is actually a composite one-step method, but the historical name has stuck.) First, intermediate solutions are computed at $(n + \frac{1}{2}, i \pm \frac{1}{2})$ using Lax-Friedrichs:

$$w_{i+\frac{1}{2}}^{n+\frac{1}{2}} = \frac{1}{2}(w_i^n + w_{i+1}^n) - \frac{\Delta t}{2\Delta x}(f_{i+1}^n - f_i^n) \qquad (8.65)$$

$$w_{i-\frac{1}{2}}^{n+\frac{1}{2}} = \frac{1}{2}(w_{i-1}^n + w_i^n) - \frac{\Delta t}{2\Delta x}(f_i^n - f_{i-1}^n). \qquad (8.66)$$

Then the new solution at $(n + 1, i)$ is computed from these two intermediate solutions using leapfrog:

$$w_i^{n+1} = w_i^n - \frac{\Delta t}{\Delta x}\left(f_{i+\frac{1}{2}}^{n+\frac{1}{2}} - f_{i-\frac{1}{2}}^{n+\frac{1}{2}}\right). \qquad (8.67)$$

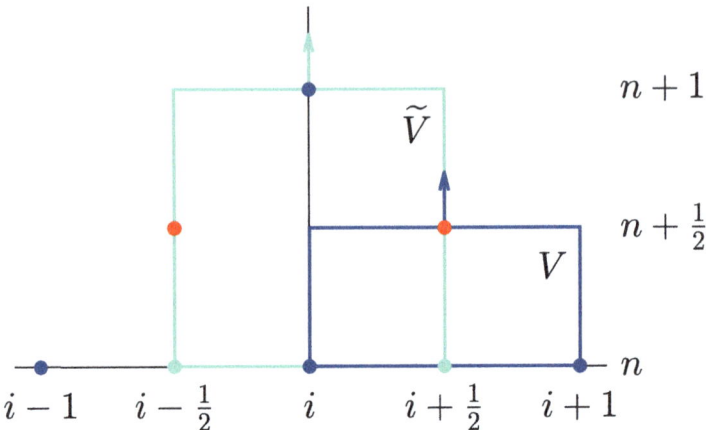

Fig. 8.5 Space-time stencil for the two-step Lax-Wendroff method. One out of the four normal vectors to V (and to $\tilde{V}$) is shown

Two-step LW is second-order accurate, stable for $\Delta t \leq \Delta x/c$, and conservative.

We will derive the two-step Lax-Wendroff method in a manifestly conservative way as a space-time finite volume method (see Fig. 8.5), using the 2D Gauss Divergence Theorem in (t, x) to discretize $w_t + f(w)_x = 0$. Write $\nabla = (\partial_t, \partial_x)$ for this derivation.

Consider first the space-time rectangle V with

$$x_i \leq x \leq x_{i+1}, \quad t_n \leq t \leq t_{n+\frac{1}{2}}. \tag{8.68}$$

Integrating over this rectangle and using the 2D Gauss Divergence Theorem implies

$$0 = \int_V (w_t + f_x) \, dt \, dx = \int_V \nabla \cdot (w, f) \, dt \, dx = \int_{\partial V} (w, f) \cdot \hat{\mathbf{n}} \, ds. \tag{8.69}$$

The four normal vectors $\hat{\mathbf{n}} = (n_t, n_x)$ are $(1, 0)$, $(0, 1)$, $(-1, 0)$, and $(0, -1)$, clockwise from the top. Approximating the integral around the perimeter ∂V of the rectangle yields

$$w_{i+\frac{1}{2}}^{n+\frac{1}{2}} \Delta x + f_{i+1}^n \frac{\Delta t}{2} - \frac{1}{2}(w_i^n + w_{i+1}^n)\Delta x - \frac{\Delta t}{2} f_i^n = 0. \tag{8.70}$$

Dividing by Δx and rearranging gives the first step of two-step LW:

$$w_{i+\frac{1}{2}}^{n+\frac{1}{2}} = \frac{1}{2}(w_i^n + w_{i+1}^n) - \frac{\Delta t}{2\Delta x}(f_{i+1}^n - f_i^n). \tag{8.71}$$

Shifting $i \to i - 1$ yields

$$w_{i-\frac{1}{2}}^{n+\frac{1}{2}} = \frac{1}{2}(w_{i-1}^n + w_i^n) - \frac{\Delta t}{2\Delta x}(f_i^n - f_{i-1}^n). \tag{8.72}$$

Thus these intermediate solutions at $(n + \frac{1}{2}, i \pm \frac{1}{2})$ are calculated from the Lax-Friedrichs method.

Next consider the space-time rectangle $\widetilde{V}$ with

$$x_{i-\frac{1}{2}} \le x \le x_{i+\frac{1}{2}}, \quad t_n \le t \le t_{n+1}. \tag{8.73}$$

Integrating over this rectangle and using the 2D Gauss Divergence Theorem implies

$$0 = \int_{\widetilde{V}} (w_t + f_x)\, dt\, dx = \int_{\widetilde{V}} \nabla \cdot (w, f)\, dt\, dx = \int_{\partial \widetilde{V}} (w, f) \cdot \hat{\mathbf{n}}\, ds. \tag{8.74}$$

Approximating the integral around the perimeter $\partial \widetilde{V}$ of the rectangle yields

$$w_i^{n+1}\Delta x + f_{i+\frac{1}{2}}^{n+\frac{1}{2}}\Delta t - w_i^n \Delta x - \Delta t f_{i-\frac{1}{2}}^{n+\frac{1}{2}} = 0. \tag{8.75}$$

Dividing by Δx and rearranging gives the second step of two-step LW:

$$w_i^{n+1} = w_i^n - \frac{\Delta t}{\Delta x}\left(f_{i+\frac{1}{2}}^{n+\frac{1}{2}} - f_{i-\frac{1}{2}}^{n+\frac{1}{2}}\right). \tag{8.76}$$

Thus the new solution at $(n + 1, i)$ is calculated from the leapfrog method.

Two-step LW is second-order accurate, CFL stable, and conservative, since its derivation is manifestly conservative.

The two-step Lax-Wendroff method is implemented in our MATLAB code **wave2.m** for the second-order wave equation, written as a two-equation first-order system (8.23) and (8.24).

8.11 Rankine-Hugoniot Jump Conditions

Nonlinear hyperbolic conservation laws support discontinuous waves, like shock waves for the inviscid Burgers equation, and shock waves and contact waves in gas dynamics, which can develop even from smooth initial data. The conserved variables obey the Rankine-Hugoniot jump conditions at these discontinuities.

For the 1D nonlinear hyperbolic conservation law

$$w_t + f(w)_x = 0 \tag{8.77}$$

where w and f have m elements in general, the jump conditions are

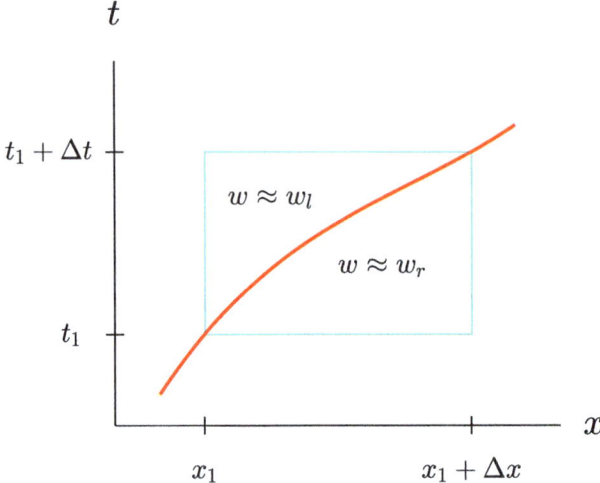

Fig. 8.6 Space-time diagram for deriving the Rankine-Hugoniot jump conditions across a discontinuity propagating with velocity $s(t)$

$$s[w] = [f] \tag{8.78}$$

where s is the discontinuity velocity and $[\chi] \equiv \chi_a - \chi_b$ with χ_a just ahead of the discontinuity and χ_b just behind. (The opposite sign convention for $[\chi]$ is equally valid.)

To derive the jump conditions (see Fig. 8.6), first note that $\Delta x \approx s(t_1)\Delta t$. Then integrating the conservation law over the space-time rectangle,

$$\int_{x_1}^{x_1+\Delta x} w(x, t_1 + \Delta t)\, dx - \int_{x_1}^{x_1+\Delta x} w(x, t_1)\, dx$$

$$= -\int_{t_1}^{t_1+\Delta t} f(w(x_1 + \Delta x, t))\, dt + \int_{t_1}^{t_1+\Delta t} f(w(x_1, t))\, dt. \tag{8.79}$$

Approximating the integrals, we obtain

$$\Delta x\, w_l - \Delta x\, w_r \approx -\Delta t\, f(w_r) + \Delta t\, f(w_l) \tag{8.80}$$

or as $\Delta t \to 0$,

$$s(w_r - w_l) = s[w] = f(w_r) - f(w_l) = [f]. \tag{8.81}$$

8.12 Solutions to Nonlinear Hyperbolic Conservation Laws

A *discontinuous solution* (or *generalized solution*) to a hyperbolic conservation law $w_t + f(w)_x = 0$ can be verified (i) by using the integral formulation

$$\int_{t_1}^{t_2} dt \int_{x_1}^{x_2} dx \, (w_t + f(w)_x) = 0 \tag{8.82}$$

or (ii) by using the (often more convenient) *weak* formulation

$$\int_0^\infty dt \int_{-\infty}^\infty dx \, (w\phi_t + f(w)\phi_x) = -\int_{-\infty}^\infty dx \, \phi(x, 0)w(x, 0) \tag{8.83}$$

of the conservation law, where $\phi(x,t)$ is a smooth function that vanishes as $t \to \infty$ for any x and as $x \to \pm\infty$ for any t, or (iii) by showing the solution satisfies the conservation law piecewise in smooth regions and the Rankine-Hugoniot jump conditions $s[w] = [f]$ at discontinuities, where s is the discontinuity velocity. We will opt for (iii) below.

A solution is verified as a *physical solution* by showing it satisfies an entropy condition or is the vanishing viscosity limit solution or is stable against small perturbations (investigate how small perturbations are transported along characteristics).

The Lax-Wendroff Theorem guarantees that if a conservative numerical method converges under spatial/temporal mesh refinement, it converges to a weak solution of the conservation law.

Theorem *Lax-Wendroff Theorem: Index a sequence of grids by $l = 1, 2, \ldots$ with $\Delta t_l \to 0$ and $\Delta x_l \to 0$ as $l \to \infty$. Denote by u_l the numerical approximation computed on the lth grid with a consistent and conservative numerical method. If $u_l(x,t) \to u(x,t)$ as $l \to \infty$, then $u(x,t)$ is a weak solution of the conservation law.*

An illuminating proof of this theorem is given in [26]. Note that a nonconservative method may converge to a function that is not a weak solution of the original PDE. Even a conservative method may require an entropy condition to converge to the correct physical solution. A good physical criterion was suggested by Lax:

Principle *Lax Entropy Condition: For a convex ($f''(w) > 0$) scalar hyperbolic conservation law $w_t + f(w)_x = 0$, a discontinuity propagating with velocity s is admissible only if $f'(w_l) > s > f'(w_r)$, where $s = [f]/[w]$. This condition implies characteristics must flow into a shock from both sides.*

8.13 Riemann Problem for Nonlinear Hyperbolic Conservation Laws

The Riemann problem for a nonlinear hyperbolic conservation law is the 1D initial value problem with constant data to the left and to the right of the origin:

$$w_t + f(w)_x = 0, \quad w(x, t = 0) = \begin{cases} w_l & x < 0 \\ w_r & x > 0 \end{cases} \quad (8.84)$$

The Riemann problem solution will be constant along rays $x/t = const$ through the origin in the space-time plane, since the conservation law and the initial data are invariant under the transformation $x \to Lx$, $t \to Lt$ for any $L > 0$. Thus, $w(x, t) = w(Lx, Lt) = w(x/t)$.

We will discuss below solutions to the Riemann problem for the inviscid Burgers equation (Sect. 8.14) and for gas dynamics (Sect. 8.19). Godunov methods (Sect. 8.27) employ (exact or approximate) Riemann problem solutions in computing solutions.

The Riemann problem solution for the inviscid Burgers equation consists of a single wave emanating from the origin, which can be either a (discontinuous) shock wave (see Fig. 8.7) or a (smooth) rarefaction wave (see Fig. 8.8), separating the original left and right states.

For polytropic gas dynamics, there are three waves emanating from the origin that separate the space-time plane into four constant-state wedges (plus any nonconstant-state rarefaction wedges): from the left, a left wave, a (discontinuous) contact wave, and then a right wave, where the left wave and the right wave may be either a (discontinuous) shock wave or a (smooth) rarefaction wave (look ahead to Fig. 8.16).

Note that for both the inviscid Burgers equation and for gas dynamics, the shock wave is an exact discontinuity, while the rarefaction wave varies smoothly through a wedge with finite width in the space-time plane. Also note that any of the waves in the Riemann problem solution can be of zero strength.

8.14 Inviscid Burgers Equation and Riemann Problem

The (viscous) Burgers equation

$$u_t + u u_x = v u_{xx} \quad (8.85)$$

is the simplest PDE that models the more complicated Navier-Stokes equations (viscous fluid dynamics, boundary layers, etc.). Here $u(x, t)$ is the fluid velocity and v is the viscosity. The viscous Burgers equation is parabolic and can be solved numerically with TRBDF2, and central differences (the advection term $u u_x$ can be

8.14 Inviscid Burgers Equation and Riemann Problem

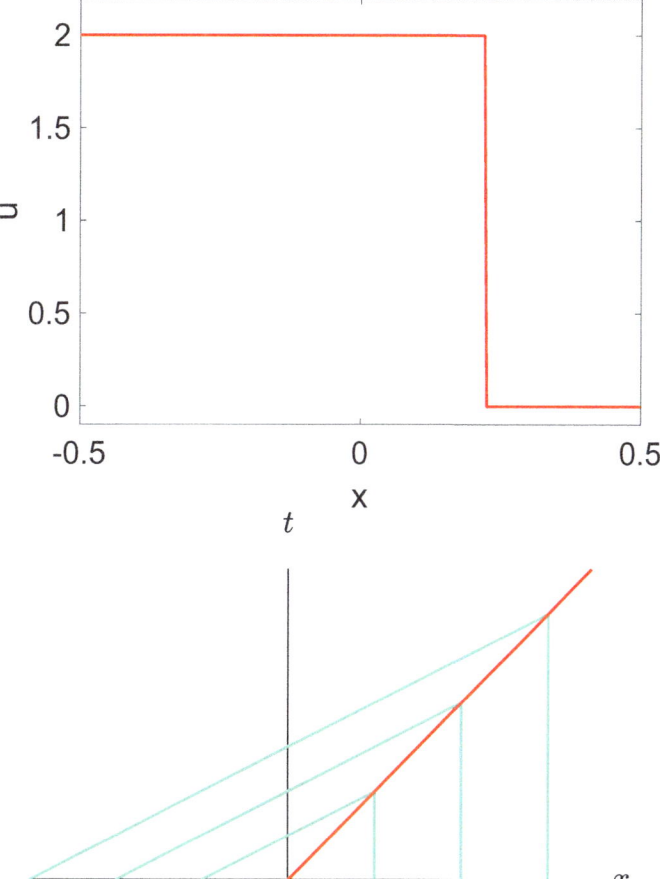

Fig. 8.7 Shock wave solution (propagating to the right) of the Riemann problem for the inviscid Burgers equation when $u_l > u_r$: u vs. x and in the x-t plane. The characteristics (*light lines*) disappear into the discontinuous shock, preventing the solution from becoming multivalued. Here $u_l = 2$, $u_r = 0$, and the shock velocity $s = 1$

discretized with upwinding, though upwinding is only required if $||uu_x|| \gg ||vu_{xx}||$ in some regions).

The *inviscid* Burgers equation

$$u_t + uu_x = 0 \tag{8.86}$$

is the simplest PDE that models the more complicated equations of gas dynamics (shock and rarefaction waves). Note that the inviscid Burgers equation can be written in conservation law form as

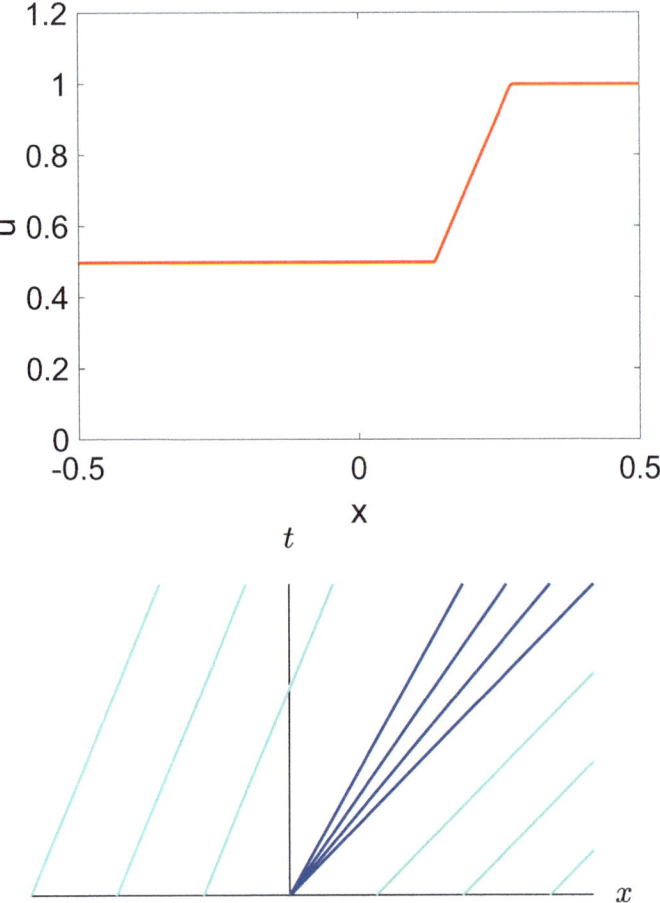

Fig. 8.8 Rarefaction wave solution (propagating to the right) of the Riemann problem for the inviscid Burgers equation when $u_l < u_r$: u vs. x and characteristic fan (*dark lines*) emanating from the origin in the x-t plane, with the exterior characteristics (*light lines*) parallel to the outer envelope of the rarefaction. Here $u_l = \frac{1}{2}$ and $u_r = 1$

$$u_t + f(u)_x = 0, \quad f(u) = \frac{1}{2}u^2. \tag{8.87}$$

The hyperbolic inviscid Burgers equation (8.86) is the vanishing viscosity ($\nu \to 0$) limit of the parabolic (viscous) Burgers equation (8.85).

Shock and rarefaction waves for the inviscid Burgers equation can be understood from the theory of characteristics. There is one family of characteristics with characteristic velocity u. The solution u is constant along these characteristics λ with $x = ut + x_0$:

8.14 Inviscid Burgers Equation and Riemann Problem

$$\left.\frac{du}{dt}\right|_\lambda = \frac{\partial u}{\partial t} + \frac{dx}{dt}\frac{\partial u}{\partial x} = \frac{\partial u}{\partial t} + u\frac{\partial u}{\partial x} = 0. \tag{8.88}$$

The inviscid Burgers equation $u_t + uu_x = 0$ with Riemann problem initial data ($u_l > u_r$)

$$u_0(x) = \begin{cases} u_l & x < 0 \\ u_r & x > 0 \end{cases} \tag{8.89}$$

has the stable, physical shock wave solution (see Fig. 8.7)

$$u(x,t) = u(x/t) = \begin{cases} u_l & x < st \\ u_r & x > st \end{cases} = u_0(x - st) \tag{8.90}$$

with shock velocity $s = \frac{1}{2}(u_l + u_r)$. This is a solution, since constants u_l and u_r solve the inviscid Burgers equation, and the shock discontinuity satisfies the Rankine-Hugoniot jump condition

$$s[u] = \left[\frac{1}{2}u^2\right]. \tag{8.91}$$

The shock wave solution is stable and physical, since it satisfies the Lax entropy condition that characteristics must flow into a shock wave from both sides (see Sect. 8.12). Imagine making a small perturbation near the shock. This perturbation then flows along characteristics into the shock and disappears, demonstrating that the shock is stable against small perturbations.

The shock wave solution is also the vanishing viscosity limit of the viscous shock profile (Fig. 8.9) traveling wave solution (see Problem 8.10) of Burgers' equation

$$u(x,t) = w(x - st) \tag{8.92}$$

where $s = (u_l + u_r)/2$ and

$$w(y) = u_r + \frac{1}{2}(u_l - u_r)\left[1 - \tanh\frac{(u_l - u_r)y}{4\nu}\right], \quad u_l > u_r. \tag{8.93}$$

If $u_l < u_r$, the stable, physical solution to the Riemann problem for the inviscid Burgers equation is a rarefaction wave (see Fig. 8.8):

$$u(x,t) = u(x/t) = \begin{cases} u_l & x < u_l t \\ x/t & u_l t \leq x \leq u_r t \\ u_r & x > u_r t \end{cases} \tag{8.94}$$

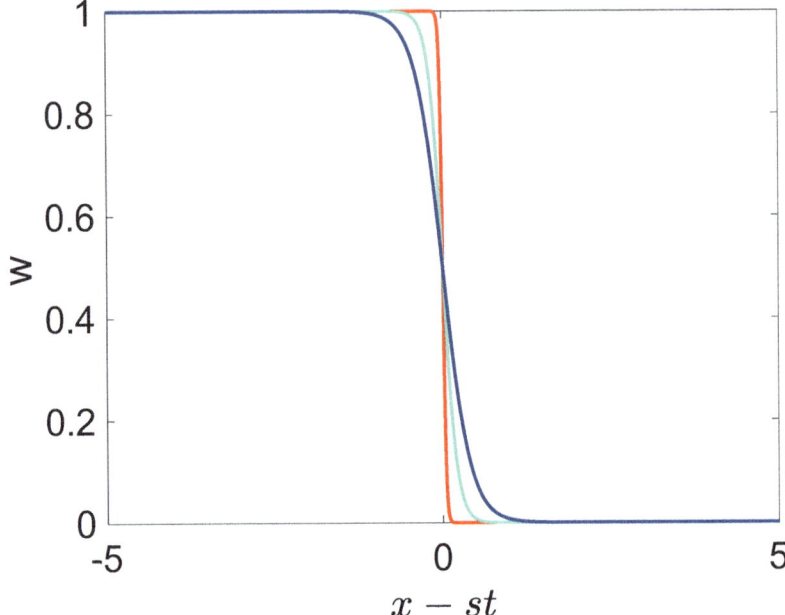

Fig. 8.9 Analytical traveling wave (viscous shock profile) solutions in the frame of the shock to Burgers' equation with $\nu = 0.01$ (*steepest curve*), 0.05 (*smoother*), and 0.1 (*smoothest*)

Now imagine making a small perturbation in the rarefaction wave. This perturbation then flows along a characteristic within the rarefaction, demonstrating stability of the wave against small perturbations.

N-waves (shock/rarefaction/shock) are transients and decay $\sim 1/\sqrt{t}$. They are next to leading order in the large t asymptotics. The leading order equals the net difference in total shock strength minus total rarefaction strength, which does not decay.

The N-wave for the inviscid Burgers equation (see Fig. 8.10) is

$$u(x,t) = \begin{cases} x/t & -\sqrt{t} \leq x \leq \sqrt{t} \\ 0 & \text{otherwise} \end{cases} \tag{8.95}$$

To show $u(x,t)$ solves the inviscid Burgers equation, note that x/t and 0 are piecewise solutions, and the Rankine-Hugoniot jump relations are satisfied at each shock wave ($x = \pm\sqrt{t}$, $u = \pm 1/\sqrt{t}$) with shock velocity $s = \pm 1/(2\sqrt{t})$, respectively:

$$s[u] = s\frac{\pm 1}{\sqrt{t}} = \frac{1}{2t} = \left[\frac{1}{2}u^2\right]. \tag{8.96}$$

8.15 Numerical Methods for the Inviscid Burgers Equation

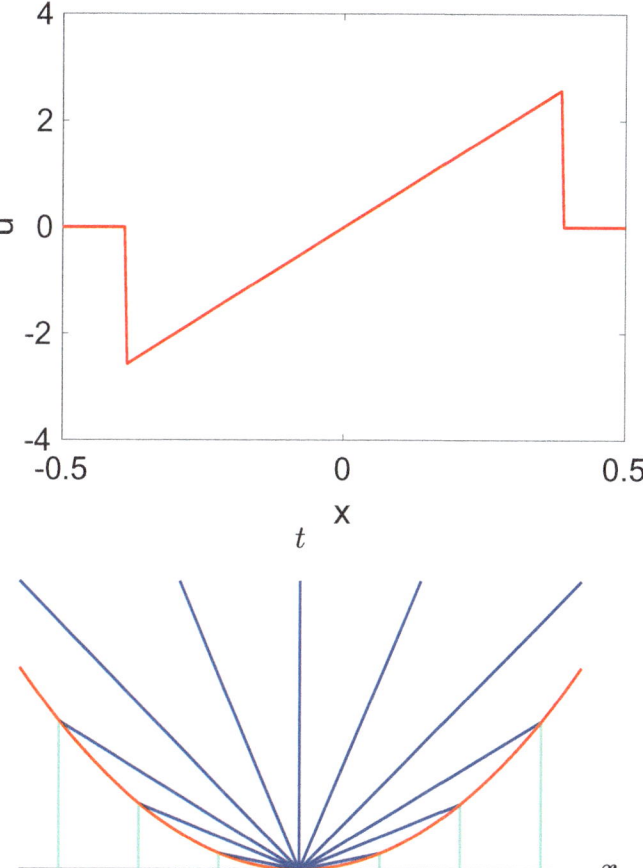

Fig. 8.10 N-wave solution u vs. x and shocks (*parabolic curve*) plus rarefaction fan (*dark lines*) emanating from the origin in the x-t plane for the inviscid Burgers equation. The left shock is propagating to the left and the right shock to the right. The characteristics (*light and dark lines*) disappear into the shocks

For our Burgers equation programs, we will use through-flow boundary conditions (see Sect. 8.20.2).

8.15 Numerical Methods for the Inviscid Burgers Equation

The conservative first-order upwind method works well for the inviscid Burgers equation

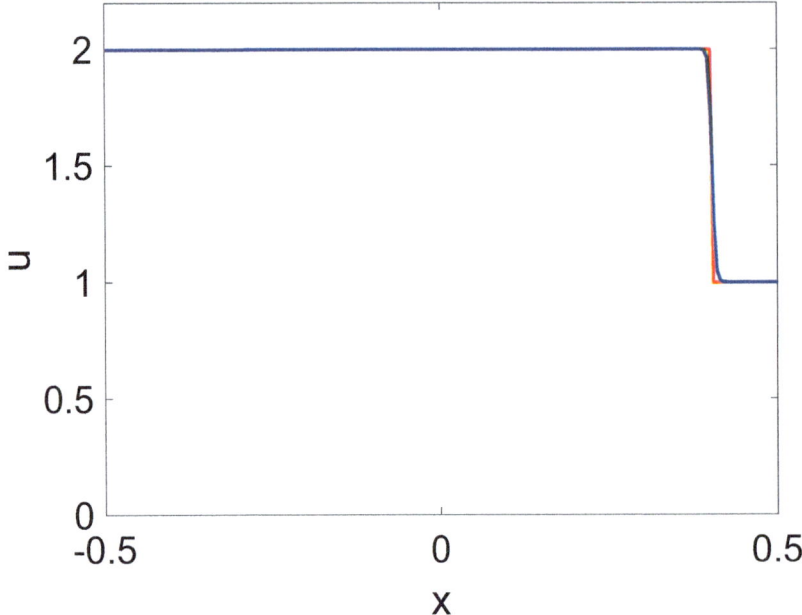

Fig. 8.11 Inviscid Burgers' equation shock wave (propagating to the right with speed $s = 1.5$) computed with the conservative upwind method using **burgers1.m** with $r = 0.9$ and $200\Delta x$ at $t = 0.27$ compared with the exact solution (*sharper curve*). Similar results are obtained with the best Lax-Friedrichs method (**burgers1.m**) and WENO3 (**burgers3.m**). Even better results are obtained with $400\Delta x$

$$u_t + \left(\frac{1}{2}u^2\right)_x = 0. \tag{8.97}$$

First let's assume all $u(x, t) > 0$. Then the conservative first-order upwind method is

$$u_i^{n+1} = u_i^n - \frac{\Delta t}{\Delta x}\left(\frac{1}{2}(u_i^n)^2 - \frac{1}{2}(u_{i-1}^n)^2\right). \tag{8.98}$$

A shock wave solution is illustrated in Fig. 8.11. Note the excellent agreement between the computed and exact solutions. If a *nonconservative* upwind discretization of the non-conservation law form $u_t + uu_x = 0$ is used,

$$u_i^{n+1} = u_i^n - \frac{\Delta t}{\Delta x}u_i^n(u_i^n - u_{i-1}^n) \tag{8.99}$$

then the computed shock velocity (see Fig. 8.12) shows a large discrepancy from the exact shock velocity. A nonconservative scheme can also produce unphysical

8.15 Numerical Methods for the Inviscid Burgers Equation

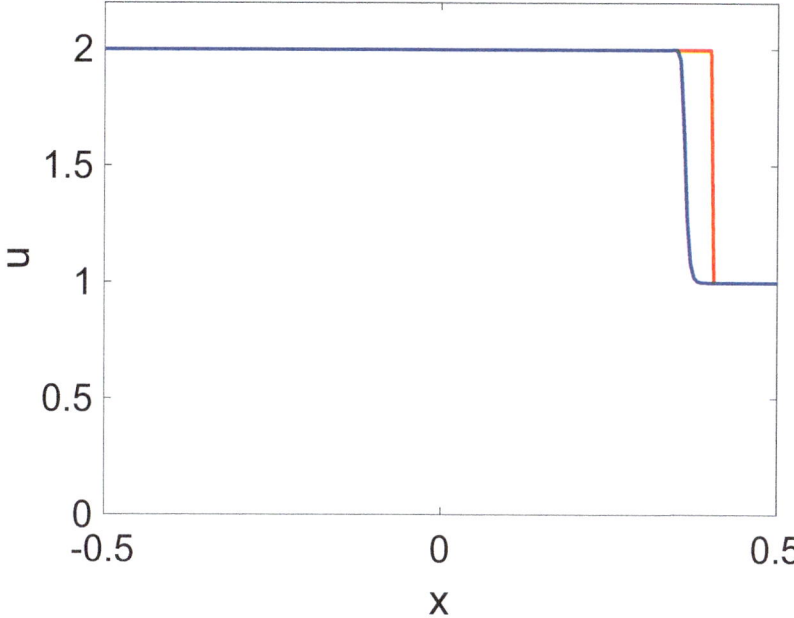

Fig. 8.12 Inviscid Burgers equation shock wave (propagating to the right) computed using **burgers1.m** with the nonconservative upwind method with $r = 0.9$ and $200\Delta x$ at $t = 0.27$ compared with the exact solution (*sharper curve*), illustrating incorrect shock speed

waves: see **burgers1.m** for an example of an unphysical rarefaction (backward shock) wave.

Allowing now positive and negative values of $u(x, t)$, the conservative first-order upwind method for the inviscid Burgers equation is written in conservative form as

$$u_i^{n+1} = u_i^n - \frac{\Delta t}{\Delta x}\left(F_{i+\frac{1}{2}}^n - F_{i-\frac{1}{2}}^n\right) \tag{8.100}$$

where

$$F_{i+\frac{1}{2}}^n = \begin{cases} \frac{1}{2}(u_i^n)^2 & \text{if } u_{i+\frac{1}{2}}^n \geq 0 \\ \frac{1}{2}(u_{i+1}^n)^2 & \text{if } u_{i+\frac{1}{2}}^n < 0 \end{cases}, \quad F_{i-\frac{1}{2}}^n = \begin{cases} \frac{1}{2}(u_{i-1}^n)^2 & \text{if } u_{i-\frac{1}{2}}^n \geq 0 \\ \frac{1}{2}(u_i^n)^2 & \text{if } u_{i-\frac{1}{2}}^n < 0 \end{cases} \tag{8.101}$$

Note that to be consistent, $F_{i-\frac{1}{2}}^n$ must be derivable from $F_{i+\frac{1}{2}}^n$ by taking $i \to i-1$.

There are four different versions of Lax-Friedrichs, all implemented for the inviscid Burgers equation in **burgers1.m**. All four versions of Lax-Friedrichs are CFL stable, first-order accurate, diffusive, and conservative (see Problems 8.1–8.3). The original method is

$$u_i^{n+1} = \frac{1}{2}(u_{i-1}^n + u_{i+1}^n) - \frac{\Delta t}{2\Delta x}\left(\frac{1}{2}(u_{i+1}^n)^2 - \frac{1}{2}(u_{i-1}^n)^2\right) \tag{8.102}$$

while much more accurate results are obtained with the best version of the Lax-Friedrichs method:

$$u_i^{n+1} = \frac{1}{2}\left(r_{i-\frac{1}{2}}u_{i-1}^n + \left(2 - r_{i-\frac{1}{2}} - r_{i+\frac{1}{2}}\right)u_i^n + r_{i+\frac{1}{2}}u_{i+1}^n\right)$$
$$- \frac{\Delta t}{2\Delta x}\left(\frac{1}{2}(u_{i+1}^n)^2 - \frac{1}{2}(u_{i-1}^n)^2\right) \tag{8.103}$$

where

$$r_{i\pm\frac{1}{2}} = \left|u_{i\pm\frac{1}{2}}^n\right|\frac{\Delta t}{\Delta x}. \tag{8.104}$$

Note that the four different versions of Lax-Friedrichs can be generated from (8.103) by choosing the appropriate values for $r_{i\pm\frac{1}{2}}$ in place of (8.104):

$$r_{i\pm\frac{1}{2}} = 1 \quad \text{(original LF)}$$

$$r_{i\pm\frac{1}{2}} = |u|_{\max}\frac{\Delta t}{\Delta x} \quad \text{(third best LF)}$$

$$r_{i\pm\frac{1}{2}} = \max\{|u_{i\pm 1}|, |u_i|\}\frac{\Delta t}{\Delta x} \quad \text{(second best LF)} \tag{8.105}$$

where $|u|_{\max}$ is the maximum value of $|u|$ on the grid.

Problem 8.3 develops the numerical flux functions for the four different versions of Lax-Friedrichs (showing they are all conservative), and Problem 8.14 compares numerical results for the four different versions of Lax-Friedrichs plus the conservative upwind method for a shock wave, rarefaction wave, merging shocks, and N-wave.

Interestingly, for the shock wave case, the conservative first-order upwind and the best version of Lax-Friedrichs give comparable results to WENO3. The conservative first-order upwind method is used in Figs. 8.13 and 8.14 to compute rarefaction wave and N-wave solutions, respectively.

The conservative upwind method for the inviscid Burgers equation is implemented in our MATLAB code **burgers1.m**.

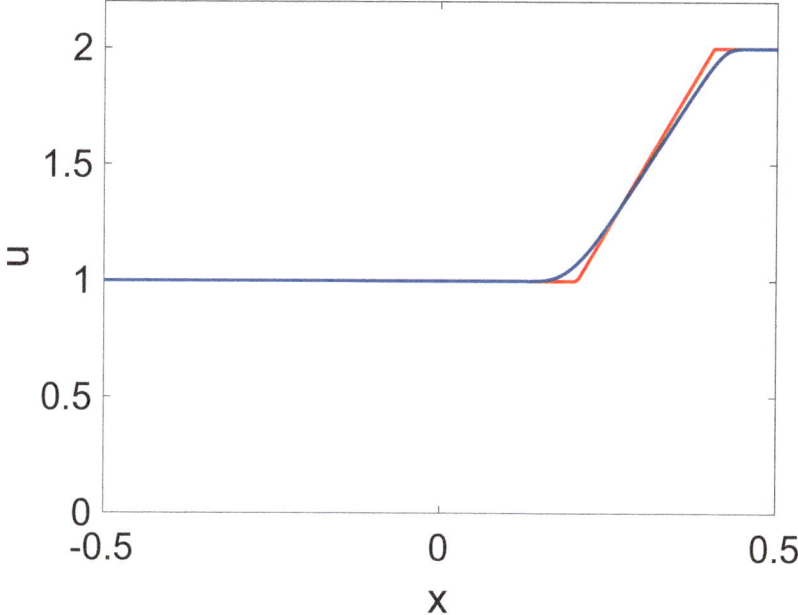

Fig. 8.13 Inviscid Burgers equation rarefaction wave (propagating to the right) computed using **burgers1.m** with the conservative upwind method with $r = 0.9$ and $200\Delta x$ at $t = 0.2$ compared with the exact solution (*sharper curve*)

8.16 Ideal Gas Dynamics

The *Euler equations of ideal gas dynamics*[1] consist of nonlinear hyperbolic conservation laws expressing conservation of mass (8.106), conservation of momentum (8.107), and conservation of energy (8.108):

$$\frac{\partial \rho}{\partial t} + \frac{\partial}{\partial x_i}(\rho u_i) = 0 \qquad (8.106)$$

$$\frac{\partial}{\partial t}(\rho u_j) + \frac{\partial}{\partial x_i}(\rho u_i u_j + P\delta_{ij}) = 0 \qquad (8.107)$$

$$\frac{\partial E}{\partial t} + \frac{\partial}{\partial x_i}(u_i(E + P)) = 0 \qquad (8.108)$$

where ρ is the mass density of the gas, **u** is the velocity, P is pressure, and E is the energy density. Indices i, j equal 1, 2, 3, and repeated indices are summed over.

---

[1] Neglect viscosity and heat conduction in the compressible Navier-Stokes equations in Sect. 9.2.

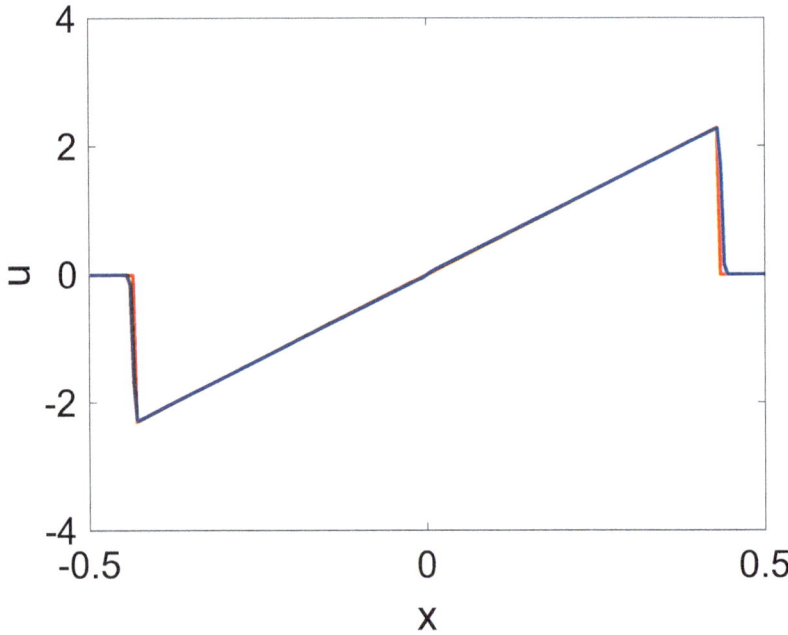

Fig. 8.14 Inviscid Burgers equation N-wave computed using **burgers1.m** with the conservative upwind method with $r = 0.9$ and $200\Delta x$ at $t = 0.125$ compared with the exact solution (*sharper curve*). The left shock is propagating to the left and the right shock to the right

The five conservation laws (8.106)–(8.108) have six unknowns: closure is provided by an equation of state that specifies the pressure P in terms of the other state variables ρ, **u**, and E. For the most widely applicable case of a polytropic (γ-*law*) gas, the pressure P is given by the equation of state

$$P = (\gamma - 1)\left(E - \frac{1}{2}\rho u^2\right). \tag{8.109}$$

Classical theory predicts $\gamma = 1 + 2/f$, where f is the number of degrees of freedom/molecule: $\gamma = 5/3$ for monatomic gases like H, He, and Ar; $\gamma = 7/5$ for diatomic gases with center of mass plus rotational modes like H_2, O_2, and N_2; and $\gamma = 9/7 \approx 1.3$ for diatomic gases with center of mass plus rotational plus vibrational modes like H_2O, CO_2, and NH_3. For air (approximately 80% N_2 and 20% O_2), $\gamma = 7/5$; for astrophysical flows (dominated by H), γ is typically set to 5/3. Note that in the real world, γ will depend on temperature and will be roughly constant only in a certain temperature range.

For an ideal gas, the temperature T is related to the pressure through the ideal gas law $P = nk_BT$, where the number density $n = \rho/m$, m is the mass of the gas particles, and k_B is Boltzmann's constant.

8.17 Gas Dynamical Shocks and Contacts

The conserved variables—always to be used in computations—are mass density, momentum density, and energy density:

$$\begin{bmatrix} \rho \\ \rho \mathbf{u} \\ E \end{bmatrix}. \tag{8.110}$$

A single variable name like **p** can be used for the momentum density $\rho \mathbf{u}$, but care must be taken to always distinguish it from the pressure, which is often written as lower case p.

The conservation laws (8.106)–(8.108) are compactly expressed as

$$q_t + f(q)_x + g(q)_y + h(q)_z = 0 \tag{8.111}$$

where the fluxes are

$$f(q) = \begin{bmatrix} \rho u \\ \rho u^2 + P \\ \rho uv \\ \rho uw \\ u(E+P) \end{bmatrix}, \quad g(q) = \begin{bmatrix} \rho v \\ \rho uv \\ \rho v^2 + P \\ \rho vw \\ v(E+P) \end{bmatrix}, \quad h(q) = \begin{bmatrix} \rho w \\ \rho uw \\ \rho vw \\ \rho w^2 + P \\ w(E+P) \end{bmatrix} \tag{8.112}$$

with $\mathbf{u} = (u, v, w)$.

8.17 Gas Dynamical Shocks and Contacts

When analyzing a set of nonlinear hyperbolic conservation laws, the possible discontinuous waves should be classified, as well as the families of characteristics and their Riemann invariants (see the next Sect. 8.18).

Let's examine 3D plane wave surfaces of discontinuity propagating along the x direction, with all $\partial_y \equiv 0$ and $\partial_z \equiv 0$. In the frame of the plane wave, where the discontinuity velocity $s^{(\text{comoving})} = 0$, the Rankine-Hugoniot jump conditions are

$$[\rho u] = 0 \quad \text{(mass flux)} \tag{8.113}$$

$$[\rho u^2 + P] = 0 \quad (x \text{ momentum flux}) \tag{8.114}$$

$$[\rho uv] = 0 \quad (y \text{ momentum flux}) \tag{8.115}$$

$$[\rho uw] = 0 \quad (z \text{ momentum flux}) \tag{8.116}$$

$$[u(E+P)] = 0 \quad \text{(energy flux)} \tag{8.117}$$

Fig. 8.15 Plane-wave discontinuity types in 3D gas dynamics depicted in the frame comoving with the discontinuity, where "a" designates ahead of the discontinuity and "b" behind the discontinuity. Here s_a and s_b are the entropies ahead of and behind the shock, respectively

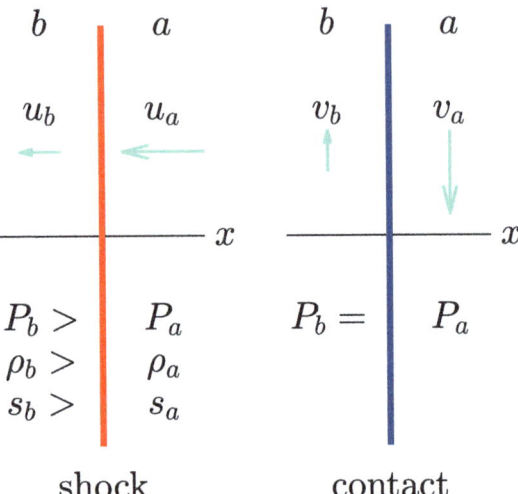

where the 3D velocity is (u, v, w).

There are two discontinuous surface types, classified by whether the mass flux across the surface is zero or not (see Fig. 8.15):

(i) *Contact wave:* mass flux = 0 (like the boundary in an ocean wave between water and air). In this case $u_a = 0 = u_b$, where a labels ahead of the wave and b labels behind the wave. The jump condition (8.114) then reduces to $[P] = 0$. Thus, at a contact wave (tangential discontinuity), the normal velocity[2] u and the pressure P are continuous, while the density ρ, tangential velocity (v, w), and entropy s can have arbitrary jumps.

(ii) *Shock wave:* mass flux $\neq 0$ (like around a supersonic aircraft). In this case, $u_a \neq 0$ and $u_b \neq 0$. Equations (8.115) and (8.116) imply $[v] = 0$ and $[w] = 0$. It is then simplest to view the shock wave in a frame where $v = 0$ and $w = 0$. Thus at a shock wave, ρ, P, and u are discontinuous, while the tangential velocity (v, w) is continuous. For a physical shock wave where the entropy is higher behind the shock, P and ρ are higher behind the shock as well, and $|u_b| < |u_a|$ (both u_a and u_b are negative in the frame of the shock; if in the lab frame the gas ahead of the shock is at rest, then $s^{(\text{lab})} = -u_a > 0$ and $s^{(\text{lab})} > u_b^{(\text{lab})} = u_b + s^{(\text{lab})} > 0$).

To describe jumps across a physical shock wave, let's start with the property that physical shock waves are compressive: $\rho_b > \rho_a$. This then implies $P_b > P_a$ from (8.114):

$$P_b - P_a = j(u_a - u_b) > 0 \tag{8.118}$$

---

[2] When we switch back to the lab frame where the discontinuity is moving at velocity $s^{(\text{lab})} \neq 0$, $u_a^{(\text{lab})} = s^{(\text{lab})} = u_b^{(\text{lab})}$.

using $j = \rho_b u_b = \rho_a u_a = const < 0$ in the frame of the shock.

It is straightforward to derive from the Rankine-Hugoniot jump relations that

$$e_b - e_a = \frac{P_a + P_b}{2}\left(\frac{1}{\rho_a} - \frac{1}{\rho_b}\right) \tag{8.119}$$

where the energy density $E = \rho e + \frac{1}{2}\rho u^2$ and e is the specific internal energy, and the relation governing the pressure jump P_b/P_a across the shock

$$\left(\frac{1}{\rho_b} - \frac{\mu^2}{\rho_a}\right) P_b = \left(\frac{1}{\rho_a} - \frac{\mu^2}{\rho_b}\right) P_a \tag{8.120}$$

where $\mu^2 = (\gamma - 1)/(\gamma + 1)$ (see Problem 8.20). The limiting strong shock density jump is obtained by taking $P_b/P_a \to \infty$: $\rho_b/\rho_a = 1/\mu^2$. For a monatomic gas like H with $\gamma = 5/3$, the limiting strong shock density jump is $\rho_b/\rho_a = 4$, while for air with $\gamma = 7/5$, the limiting $\rho_b/\rho_a = 6$. Much stronger density contrasts are attained in real-world situations like astrophysical shocks with radiative cooling, where the density ratio can be arbitrarily high.

The entropy of a polytropic gas is $s = c_V \ln(P/\rho^\gamma)$, where c_V is the specific heat capacity at constant volume. To show the entropy increases behind the shock wave, we verify using (8.120) that

$$\frac{s_b - s_a}{c_V} = \ln\left[\frac{P_b}{P_a}\left(\frac{\rho_a}{\rho_b}\right)^\gamma\right] = \ln\left[\frac{X - \mu^2}{(1 - \mu^2 X) X^\gamma}\right] > 0, \quad 1 < X < 1/\mu^2 \tag{8.121}$$

where the density ratio $X = \rho_b/\rho_a$, with $1 < \gamma$ and $0 < \mu^2 < 1$. The argument of the logarithm equals 1 in the weak shock limit $X = 1$ and then increases monotonically (see Problem 8.21) for $1 < X < 1/\mu^2$, so the inequality holds.

Note that under a Galilean transformation $u \to u + u^{\text{boost}}$, u, discontinuity velocities $s^{\text{(lab)}}$, and E change, but ρ, P, e, and entropy s are invariant. Thus (8.119), (8.120), and (8.121) are invariant with respect to Galilean transformations.

8.18 Characteristic Equations for Gas Dynamics

Entropy is created by shock waves. Away from discontinuities, though, the entropy s is constant (adiabatic flow) along particle paths and satisfies $ds/dt = s_t + u s_x = 0$ (see Problem 8.16).[3] Thus away from discontinuities, the 1D Euler equations take the form

---

[3] The *entropy density* $S = \rho s/m$ satisfies the entropy conservation law $S_t + (uS)_x = 0$, away from discontinuities.

$$\rho_t + (\rho u)_x = \rho_t + u\rho_x + \rho u_x = 0 \qquad (8.122)$$

$$(\rho u)_t + (\rho u^2 + P)_x = \rho u_t + \rho u u_x + P_x = 0 \qquad (8.123)$$

$$s_t + u s_x = 0. \qquad (8.124)$$

There are three families of characteristics: entropy characteristics with characteristic velocity $\lambda_0 = u$ and

$$ds = 0 \qquad (8.125)$$

(which follows directly from $ds/dt = 0$ along λ_0 from (8.124)), and pressure (sound wave) characteristics with characteristic velocities $\lambda_\pm = u \pm c$ and

$$dP \pm \rho c \, du = 0. \qquad (8.126)$$

To derive this last relation, multiply (8.122) by c^2 and (8.123) by $\pm c$, using $dP = c^2 d\rho$ to obtain:

$$P_t + (u \pm c) P_x \pm \rho c (u_t + (u \pm c) u_x) = 0 \qquad (8.127)$$

or, in other words,

$$\frac{dP}{dt} \pm \rho c \frac{du}{dt} = 0. \qquad (8.128)$$

The Riemann invariant for the entropy characteristics is

$$\Gamma_0 = s, \quad \lambda_0 = u. \qquad (8.129)$$

The Riemann invariants for the pressure characteristics are

$$\Gamma_\pm = u \pm \int \frac{c(\rho)}{\rho} \, d\rho, \quad \lambda_\pm = u \pm c. \qquad (8.130)$$

For a γ-law gas, $P = A\rho^\gamma$ away from discontinuities, and

$$\Gamma_\pm = u \pm \frac{2c}{\gamma - 1}. \qquad (8.131)$$

To show this, note that

$$c^2 = \frac{dP}{d\rho} = \frac{\gamma P}{\rho} = \gamma A \rho^{\gamma - 1} \qquad (8.132)$$

$$2c\,dc = (\gamma - 1)\gamma A\rho^{\gamma-2}\,d\rho = (\gamma - 1)c^2 \frac{d\rho}{\rho} \tag{8.133}$$

$$\frac{2\,dc}{\gamma - 1} = \frac{c}{\rho}\,d\rho. \tag{8.134}$$

Note *Soundspeed in air: Newton assumed sound wave propagation in air is isothermal ($P = (\rho/m)k_B T$ with $T = const$) and predicted the soundspeed $c = \sqrt{dP/d\rho} = \sqrt{P/\rho}$, but it is actually a reversible adiabatic process (isentropic), which implies $P = A\rho^\gamma$ with $c = \sqrt{\gamma P/\rho}$. Adiabatic means no gain or loss of heat, so reversible adiabatic implies the change in heat $dQ_{heat} = Tds = 0$, where s is the entropy. (An irreversible process would have $dQ_{heat} < Tds$.)*

8.19 Riemann Problem for Gas Dynamics

The Riemann problem for gas dynamics with initial conditions for density, velocity, and pressure

$$(\rho_l, u_l, P_l) \text{ for } x < 0, \quad (\rho_r, u_r, P_r) \text{ for } x > 0 \tag{8.135}$$

has a unique solution for a polytropic gas, consisting of three waves emanating from the origin (Fig. 8.16): from the left, a left wave L, a contact wave C at which the density jumps but the velocity and pressure are continuous, and then a right wave R, where the left wave and the right wave may be either a shock wave at which the density, velocity, and pressure jump, or a smooth rarefaction wave. Note that in general any of the three waves may be of zero strength.

The Riemann problem solution is constructed (see [7]) by using the Rankine-Hugoniot jump relations across a shock and using characteristics, Riemann invariants, and the adiabatic law[4] $P = const\,\rho^\gamma$ across a rarefaction. These conditions produce an equation for u_* from the left data and from the right data

$$u_*(P_*, \rho_l, u_l, P_l) = u_*(P_*, \rho_r, u_r, P_r) \tag{8.136}$$

which has a unique real solution for P_* for a polytropic gas.

Riemann problem (RP) solutions will provide a test suite of exact solutions to compare with computations for our gas dynamical solver **weno3.m** (Sect. 8.24). Eight exemplary Riemann problems (six are simulated in the review article by Liska and Wendroff [27], including four from [38]) are provided in **weno3.m**:

---

[4] The adiabatic law holds in regions of smooth flow, away from discontinuities.

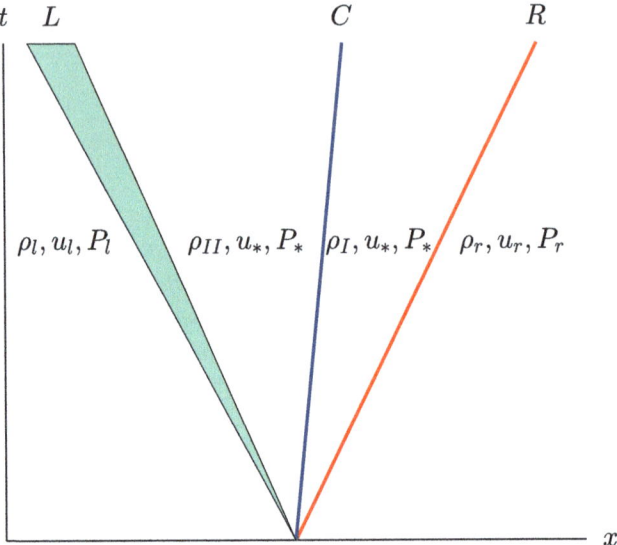

Fig. 8.16 Exact solution of a gas dynamical Riemann problem in the *x-t* plane showing (from the left) a rarefaction wave (*shaded region*), a contact wave, and a shock wave. The left wave and the right wave may be either a shock or a rarefaction. The *L–C–R* order of the waves is always preserved, but the wave velocities may be positive or negative

1. Jet RP (Ha and Gardner): shock/contact/shock
2. Sod RP: rarefaction/contact/shock
3. Toro/Sod RP: rarefaction/contact/shock
4. Stationary Contact RP (Toro): rarefaction/contact/shock
5. Near-Vacuum RP (Toro): rarefaction/contact (0 strength)/rarefaction
6. Noh RP: shock/contact (0 strength)/shock
7. Peak RP (Kucharik): rarefaction/contact/shock
8. Two Strong Shocks RP (Toro): shock/contact/shock

8.20 Boundary Conditions for Gas Dynamics

We will take gas dynamics as our paradigm for boundary conditions for hyperbolic systems $w_t + f(w)_x = 0$, but the concepts are quite general.

The best approach for using WENO for multidimensional hyperbolic problems is dimension by dimension splitting (Sect. 8.22), so we describe the ideas in 1D. For a three-point stencil for a centered second-order method like Lax-Wendroff, we need one ghost point left and right of the computational domain. For a second-order piecewise linear method like CLAWPACK, we need two ghost points left and right to implement the flux limiting. For third-order WENO, we also need two ghost

8.20 Boundary Conditions for Gas Dynamics

points left and right for the fluxes. In general, there will be m_{BC} ghost points left and right.

The simplest computational boundary conditions for hyperbolic PDEs are periodic. For bounded spatial domains, appropriate boundary conditions for hyperbolic PDEs are Cauchy (determined by how information flows along characteristics); for gas dynamics, the physically relevant Cauchy boundary conditions are *through-flow* boundary conditions or *wall* boundary conditions.

The periodic and through-flow boundary conditions are easy to implement with ghost points. The simplest no-flow wall boundary condition is easy to implement as well. But the preferred characteristic wall boundary condition is more complicated, since it involves solving characteristic equations.

Label the grid points on a spatial interval $[a, b]$ as $x_0 = a, x_1, \ldots, x_N = b$ with $N\Delta x = b - a$. The solution values are labeled $w_0, w_1, \ldots, w_N$. For MATLAB, simply shift the index $i \to i + 1$ (see Appendix A.1).

8.20.1 Periodic Boundary Conditions

With a three-point stencil, identify $w_{N+1} = w_0$ and $w_{-1} = w_N$ (see Fig. 8.17). More generally, set $w_{N+1} = w_0$, $w_{N+2} = w_1, \ldots, w_{N+m_{BC}} = w_{m_{BC}-1}$, and $w_{-1} = w_N$, $w_{-2} = w_{N-1}, \ldots, w_{-m_{BC}} = w_{N-m_{BC}+1}$.

8.20.2 Through-Flow Boundary Conditions

Zeroth-order extrapolation (see Fig. 8.18), which gives a boundary Riemann problem at $x_{-\frac{1}{2}}$ and $x_{N+\frac{1}{2}}$ with no interaction, works best here, even for high-order schemes: Set the ghost points at the left $w_{-1} = w_0, \ldots, w_{-m_{BC}} = w_0$ and the ghost points at the right $w_{N+1} = w_N, \ldots, w_{N+m_{BC}} = w_N$. Higher-order extrapolation can give rise to instabilities.

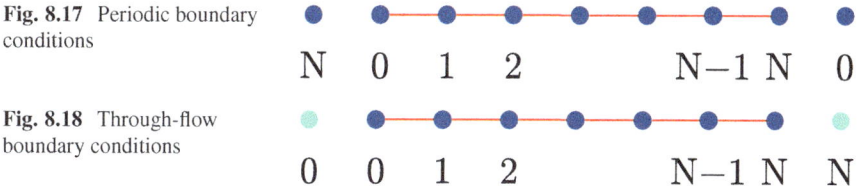

Fig. 8.17 Periodic boundary conditions

Fig. 8.18 Through-flow boundary conditions

8.20.3 Wall Boundary Conditions

At a stationary wall boundary, the normal fluid velocity u must equal 0. Label a left wall by $i = 0$ and a right wall by $i = N$. The simplest no-flow wall boundary condition sets

$$\rho_0^{n+1} = \rho_1^n, \quad \rho_0^{n+1} u_0^{n+1} = 0, \quad E_0^{n+1} = E_1^n \tag{8.137}$$

and

$$\rho_N^{n+1} = \rho_{N-1}^n, \quad \rho_N^{n+1} u_N^{n+1} = 0, \quad E_N^{n+1} = E_{N-1}^n \tag{8.138}$$

where the old and new time levels are labeled by n and $n + 1$.

A higher-quality wall boundary condition can be implemented by tracing back along characteristics that flow into the wall boundary during the timestep and solving characteristic equations (see Fig. 8.19).

At a right wall boundary, first locate the foot x_* (at time level n) of the λ_+ pressure characteristic flowing into the wall (at time level $n + 1$), estimating the characteristic velocity $\lambda_+ = u + c \approx u_N^{n+1} + c_N = c_N$ (all soundspeeds c will be evaluated at the old time level n): $x_* \approx x_N - c_N \Delta t$. Note that due to the CFL condition, $x_{N-1} \leq x_* < x_N$. Then linearly interpolate between the gas dynamical states w_N^n and w_{N-1}^n to obtain the state w_*^n. Now solve the characteristic equation $dP = -\rho c \, du$ along the λ_+ pressure characteristic to find the new pressure at the wall, approximating c by c_*:

$$P_N^{n+1} = P_*^n + \rho_*^n c_* u_*^n. \tag{8.139}$$

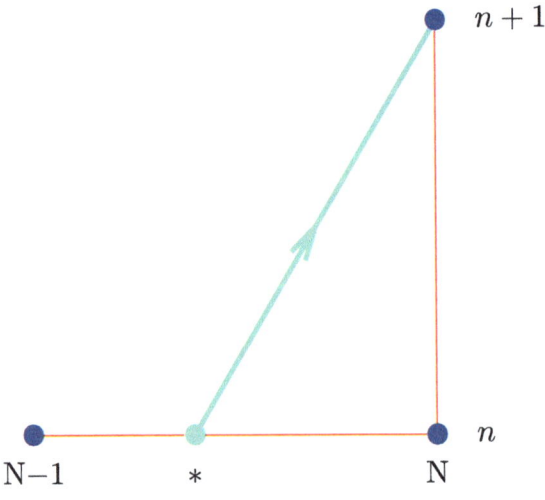

Fig. 8.19 Wall boundary condition

8.21 Space-Time Finite Volume Methods for Conservation Laws

Finally solve the soundspeed equation $d\rho = dP/c^2$ along the λ_+ pressure characteristic to obtain

$$\rho_N^{n+1} = \rho_*^n + (P_N^{n+1} - P_*^n)/c_*^2, \quad \rho_N^{n+1} u_N^{n+1} = 0, \quad E_N^{n+1} = P_N^{n+1}/(\gamma - 1). \tag{8.140}$$

Alternatively, we could solve the soundspeed equation along the entropy characteristic along the wall ($\lambda_0 = u = 0$) to find

$$\rho_N^{n+1} = \rho_N^n + (P_N^{n+1} - P_N^n)/c_N^2. \tag{8.141}$$

At a left wall boundary, first locate the foot x_* (at time level n) of the λ_- pressure characteristic flowing into the wall (at time level $n+1$), estimating the characteristic velocity $\lambda_- = u - c \approx u_0^{n+1} - c_0 = -c_0$: $x_* \approx x_0 + c_0 \Delta t$. Note that due to the CFL condition, $x_0 < x_* \leq x_1$. Then linearly interpolate between the gas dynamical states w_0^n and w_1^n to obtain the state w_*^n. Now solve the characteristic equation $dP = \rho c \, du$ along the λ_- pressure characteristic to find the new pressure at the wall, approximating c by c_*:

$$P_0^{n+1} = P_*^n - \rho_*^n c_* u_*^n. \tag{8.142}$$

Finally solve the soundspeed equation $d\rho = dP/c^2$ along the λ_- pressure characteristic to obtain

$$\rho_0^{n+1} = \rho_*^n + (P_0^{n+1} - P_*^n)/c_*^2, \quad \rho_0^{n+1} u_0^{n+1} = 0, \quad E_0^{n+1} = P_0^{n+1}/(\gamma - 1). \tag{8.143}$$

8.21 Space-Time Finite Volume Methods for Conservation Laws

Space-time finite volume methods like two-step Lax-Wendroff and Godunov's method (see Sect. 8.27) are conservative by construction. Define the cell average of $w(x, t_n)$ as

$$w_i^n = \frac{1}{\Delta x} \int_{x_{i-\frac{1}{2}}}^{x_{i+\frac{1}{2}}} w(x, t_n) \, dx. \tag{8.144}$$

Then integrate the conservation law $w_t + f(w)_x = 0$ with respect to t and x to obtain

$$\int_{x_{i-\frac{1}{2}}}^{x_{i+\frac{1}{2}}} w(x, t_{n+1}) \, dx - \int_{x_{i-\frac{1}{2}}}^{x_{i+\frac{1}{2}}} w(x, t_n) \, dx$$

$$= -\int_{t_n}^{t_{n+1}} f\left(w\left(x_{i+\frac{1}{2}},t\right)\right) dt + \int_{t_n}^{t_{n+1}} f\left(w\left(x_{i-\frac{1}{2}},t\right)\right) dt \qquad (8.145)$$

and the conservative form

$$w_i^{n+1} = w_i^n - \frac{\Delta t}{\Delta x}\left(F_{i+\frac{1}{2}} - F_{i-\frac{1}{2}}\right) \qquad (8.146)$$

where the numerical flux function

$$F_{i+\frac{1}{2}} \approx \frac{1}{\Delta t}\int_{t_n}^{t_{n+1}} f\left(w\left(x_{i+\frac{1}{2}},t\right)\right) dt. \qquad (8.147)$$

8.22 Dimension by Dimension Splitting

The finite volume method using Gauss' Divergence Theorem (see Sect. 6.12) is the method of choice for discretizing parabolic and elliptic PDEs in multiple spatial dimensions. However, it is usually too diffusive for hyperbolic problems, spreading out discontinuous waves like shock waves.[5] For WENO, dimension-by-dimension splitting gives much sharper resolution of multidimensional waves.

Write a 3D conservation law (the 2D case is exactly similar, leaving out h and z) as

$$0 = w_t + \nabla \cdot \mathbf{F}(w) = w_t + f(w)_x + g(w)_y + h(w)_z. \qquad (8.148)$$

In dimension by dimension splitting, (8.148) is simply discretized as

$$w^{n+1} = w^n - \Delta t\left(f(w^n)_x + g(w^n)_y + h(w^n)_z\right) \qquad (8.149)$$

where again the spatial discretizations are determined by the numerical method. This splitting method yields high-order accuracy for spatial derivatives without the computational cost of multidimensional finite volume reconstructions. It is the preferred splitting for WENO.

8.23 WENO Method

In this section, we will describe the third-order WENO3-LF (weighted essentially non-oscillatory with Lax-Friedrichs [flux splitting]) [33] finite difference method,

---

[5] The multidimensional finite volume method with MUSCL reconstruction can provide sharp resolution of shocks in higher-order schemes like PPM and CLAWPACK.

8.23 WENO Method

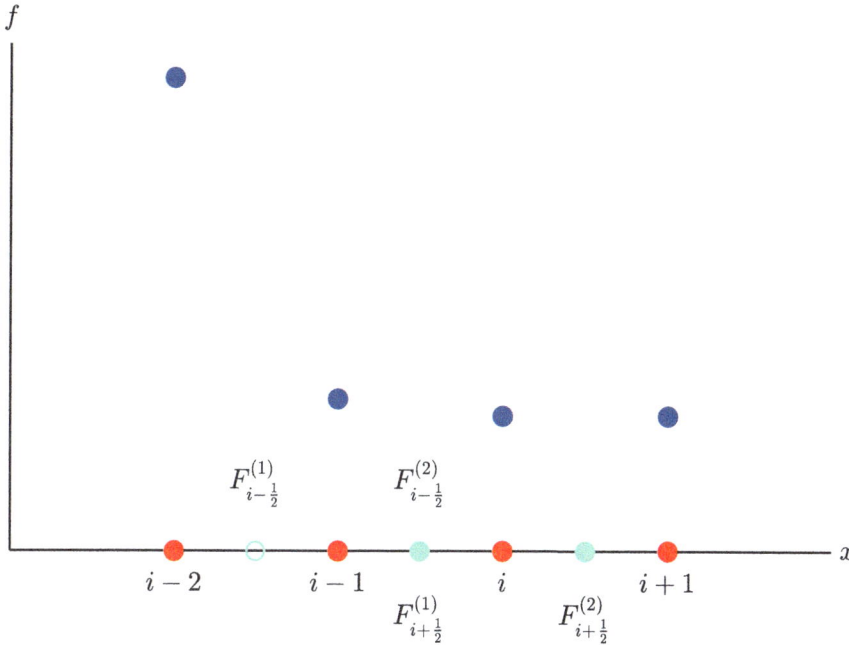

Fig. 8.20 Third-order WENO stencil for flow to the right near a shock wave (highest value of the flux f), where the contribution of the numerical flux $F^{(1)}_{i-\frac{1}{2}}$ is suppressed

which we use for simulations of gas dynamical Riemann problems, supersonic jets, and the classical hydrodynamic model. There is a plethora of other variations of ENO/WENO, including higher-order versions and finite volume, finite element, and spectral method approaches.

ENO/WENO schemes apply high-order upwind methods to nonlinear hyperbolic conservation laws with piecewise smooth solutions away from sharp discontinuities like shock waves and contacts. A nonlinear adaptive algorithm implements locally smooth stencils to avoid crossing discontinuities (see Fig. 8.20) if possible in the interpolation procedure. The weighted ENO (WENO) schemes employ a convex combination of all stencils, instead of the single stencil in the original ENO method.

To describe the WENO3 finite difference spatial discretization, we start with the scalar conservation law

$$u_t + f(u)_x = 0 \qquad (8.150)$$

and assume $\partial f/\partial u > 0$ (the wind direction is to the right). The computational domain is discretized with $N + 1$ grid points $x_i = i\Delta x$, $i = 0, 1, \ldots, N$. A conservative numerical approximation $u_i(t)$ to the exact solution $u(x_i, t)$ of (8.150) satisfies the ODE system

$$\frac{du_i(t)}{dt} + \frac{1}{\Delta x}\left(F_{i+\frac{1}{2}} - F_{i-\frac{1}{2}}\right) = 0 \tag{8.151}$$

where $F_{i+\frac{1}{2}}$ is the numerical flux function.

For third-order WENO, the numerical flux is defined as

$$F_{i+\frac{1}{2}} = \omega_1^+ F_{i+\frac{1}{2}}^{(1)} + \omega_2^+ F_{i+\frac{1}{2}}^{(2)} \tag{8.152}$$

where the two second-order accurate numerical fluxes are given by

$$F_{i+\frac{1}{2}}^{(1)} = -\frac{1}{2}f_{i-1} + \frac{3}{2}f_i, \quad F_{i+\frac{1}{2}}^{(2)} = \frac{1}{2}f_i + \frac{1}{2}f_{i+1} \tag{8.153}$$

with $f_i \equiv f(u_i(t))$. $F_{i-\frac{1}{2}}$ is constructed by shifting $i \to i-1$:

$$F_{i-\frac{1}{2}} = \omega_1^- F_{i-\frac{1}{2}}^{(1)} + \omega_2^- F_{i-\frac{1}{2}}^{(2)} \tag{8.154}$$

$$F_{i-\frac{1}{2}}^{(1)} = -\frac{1}{2}f_{i-2} + \frac{3}{2}f_{i-1}, \quad F_{i-\frac{1}{2}}^{(2)} = \frac{1}{2}f_{i-1} + \frac{1}{2}f_i. \tag{8.155}$$

Note that the stencil $\{i-2, i-1, i, i+1\}$ for (8.151) is biased to the left due to the positive wind direction.

The nonlinear weights ω_k ($k = 1, 2$) are given by

$$\omega_k^\pm = \frac{\tilde{\omega}_k^\pm}{\sum_{l=1}^2 \tilde{\omega}_l^\pm}, \quad \tilde{\omega}_l^\pm = \frac{\gamma_l}{(\epsilon + \beta_l^\pm)^2} \tag{8.156}$$

with the linear weights

$$\gamma_1 = \frac{1}{3}, \quad \gamma_2 = \frac{2}{3} \tag{8.157}$$

and the smoothness indicators

$$\beta_1^+ = (f_i - f_{i-1})^2, \quad \beta_2^+ = (f_{i+1} - f_i)^2,$$

$$\beta_1^- = (f_{i-1} - f_{i-2})^2, \quad \beta_2^- = (f_i - f_{i-1})^2. \tag{8.158}$$

The parameter ϵ insures the denominator in (8.156) does not become zero ($\epsilon = 10^{-6}$ in our computations). The magnitude of ϵ does not affect the accuracy of the simulations.

The WENO3 scheme does not invoke any free parameters. The scheme performs well for smooth solutions and for solutions containing shocks or contacts or high

8.23 WENO Method

gradient regions, because the nonlinear weights are dynamically adjusted based on local smoothness information.

If the wind direction $\partial f/\partial u < 0$, the method for computing the numerical flux $F_{i+\frac{1}{2}}$ is the exact mirror image of the scheme for $\partial f/\partial u > 0$. The stencil is now biased to the right due to the negative wind direction. If $\partial f/\partial u$ changes sign, we use a flux splitting

$$f(u) = f^+(u) + f^-(u) \tag{8.159}$$

where $\partial f^+(u)/\partial u \geq 0$ and $\partial f^-(u)/\partial u \leq 0$ and apply the method separately on f^+ and f^-. The most popular flux splitting is the Lax-Friedrichs flux splitting where

$$f^{\pm}(u) = \frac{1}{2}(f(u) \pm \alpha u) \tag{8.160}$$

with $\alpha = \max_u |\partial f(u)/\partial u|$.

For systems of hyperbolic conservation laws (8.150), the eigenvalues of the Jacobian $\partial f/\partial u$ are all real, and the Jacobian has a complete set of left and right eigenvectors. The nonlinear WENO scheme is applied in each of the local characteristic variables, obtained by using the eigenvectors of the Jacobian. To compute the numerical flux $F^n_{i+\frac{1}{2}}$ (and similarly $F^n_{i-\frac{1}{2}}$ by shifting $i \to i-1$) in (8.151), first compute the right R and left $L = R^{-1}$ Roe matrices[6] that approximate the right and left eigenvector matrices of $\partial f/\partial u$ at $i+\frac{1}{2}$. Then transform the exact fluxes $f^{\pm}$ to characteristic variables space

$$f_j^{\text{char},+} = R^{-1} f_j^+ = L f_j^+, \quad j = i-1, i, i+1 \tag{8.161}$$

$$f_j^{\text{char},-} = R^{-1} f_j^- = L f_j^-, \quad j = i, i+1, i+2 \tag{8.162}$$

for the WENO3 stencil. Now apply the WENO3 method using the $f_j^{\text{char},\pm}$ to obtain $F_{i+\frac{1}{2}}^{\text{char},\pm}$ and then transform back to physical space

$$F_{i+\frac{1}{2}}^{\text{char}} = F_{i+\frac{1}{2}}^{\text{char},+} + F_{i+\frac{1}{2}}^{\text{char},-}, \quad F_{i+\frac{1}{2}}^n = R F_{i+\frac{1}{2}}^{\text{char}}. \tag{8.163}$$

Transforming to characteristic variables to compute the fluxes allows us to simulate more difficult problems involving stronger waves. In particular for gas dynamics, if transformation to characteristic variables is not used with the WENO fluxes, the Jet, Stationary Contact, and Noh RP examples in **weno3.m** fail, while the Peak RP example gives pressure and velocity overshoots and oscillations. Similarly,

---

[6] The Roe matrices are defined with mean values at $i+\frac{1}{2}$ using u_i and u_{i+1} (see [30] and **weno3.m**).

if transformation to characteristic variables is not used for supersonic jets, the simulation below in Fig. 8.26 breaks down at Mach 7 instead of easily achieving Mach 80.

For 2D and 3D, the finite difference version of WENO simply applies WENO in each direction via dimension-by-dimension splitting (Sect. 8.22).

The temporal discretization is implemented by a third-order TVD (see Sect. 8.27) Runge-Kutta (TVD-RK3) method:

$$u^{(1)} = u^n + \Delta t \mathcal{L}(u^n, t_n) \tag{8.164}$$

$$u^{(2)} = \frac{3}{4}u^n + \frac{1}{4}u^{(1)} + \frac{1}{4}\Delta t \mathcal{L}(u^{(1)}, t_n + \Delta t) \tag{8.165}$$

$$u^{n+1} = \frac{1}{3}u^n + \frac{2}{3}u^{(2)} + \frac{2}{3}\Delta t \mathcal{L}\left(u^{(2)}, t_n + \frac{1}{2}\Delta t\right) \tag{8.166}$$

where $\mathcal{L}$ is the approximation of the spatial derivative $\mathcal{L}(u, t) \approx -f_x$ by the WENO scheme. The TVD-RK3 temporal discretization, consisting of convex combinations of three first-order forward Euler steps, is stable if the first-order forward Euler method is stable. The CFL stability condition is

$$\alpha \frac{\Delta t}{\Delta x} \equiv r \leq 1 \tag{8.167}$$

where α is the largest eigenvalue in absolute value of the Jacobian $\partial f/\partial u$. In nonlinear computations, the Courant number r is typically set between 0.1 and 0.9.

For the astrophysical jet simulations, a radiative cooling term is included in the energy conservation equation for gas dynamics

$$\frac{\partial E}{\partial t} + \cdots = -C(n, T) \tag{8.168}$$

which is incorporated within the TVD-RK3 solver. There are analogous source terms in the classical hydrodynamic model equations (see Sect. 9.5).

We highly recommend using a positivity-preserving [21] version of the WENO3 method, implemented in **weno3.m**. Most numerical methods for gas dynamics can produce negative densities and pressures during strong wave interactions. Positivity-preserving methods ensure that the gas density ρ and pressure P always remain positive by limiting the numerical flux. Without the incorporation of the positivity-preserving flux limiter, the simulations shown in the Frontispiece and Fig. 8.21 broke down as soon as the jet bow shock impacted the dense molecular clouds.

Of the classical methods for gas dynamics, only the Lax-Friedrichs method guarantees that $\rho > 0$ and $P > 0$. However, by itself Lax-Friedrichs is too diffusive to give good results for gas dynamics. Thus, we use WENO with a positivity-preserving flux limiter when needed, which can be constructed by mixing in the Lax-Friedrichs flux with the WENO flux:

8.23 WENO Method

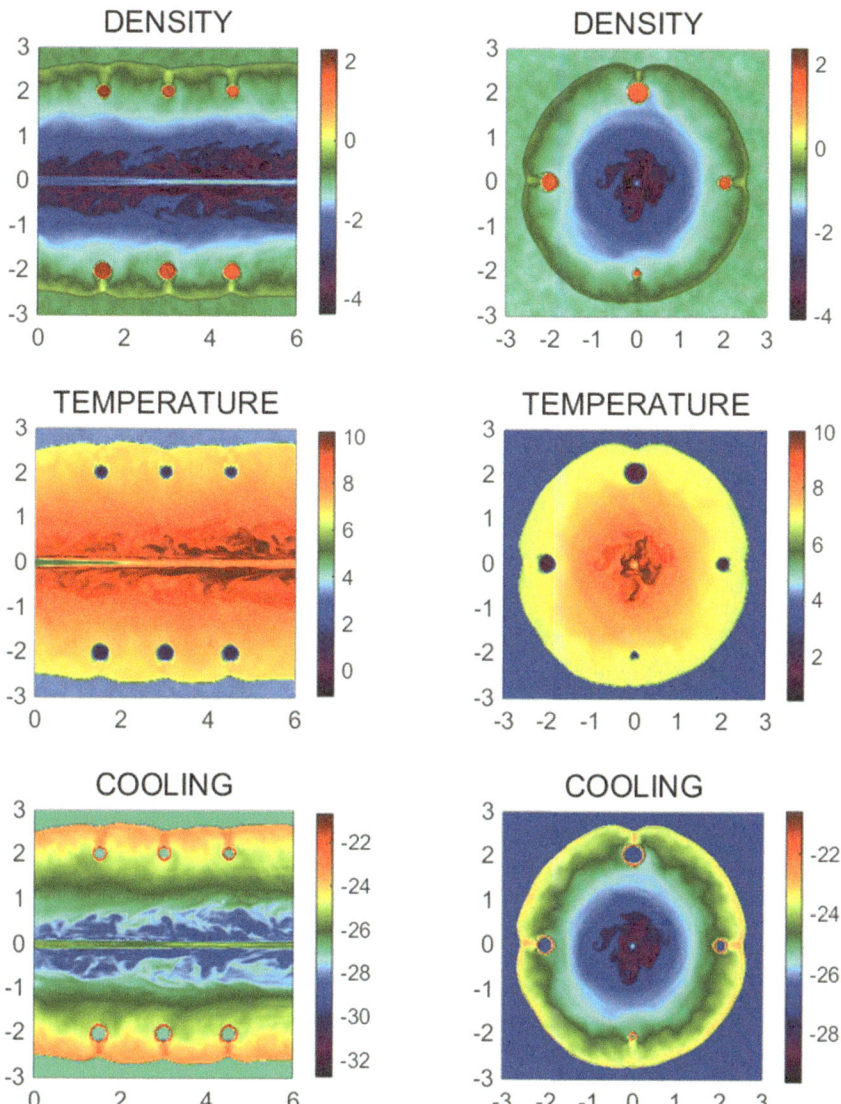

Fig. 8.21 Longitudinal and transverse cross sections of a 3D WENO3 simulation by C. L. Gardner, J. R. Jones, E. Scannapieco, and R. A. Windhorst [15] of star formation by the Centaurus A jet: $\log_{10}(n)$ with n in H/cm^3, $\log_{10}(T)$ with T in K, and $\log_{10}(C)$ with C in erg/(cm^3 s) at 780,000 year. The jet tip is propagating at Mach 1000 with respect to the ambient gas. The grid is 6300 light years on each side, with $420\Delta x \times 420\Delta y \times 420\Delta z$. The simulations took 100,000 CPU hours = 11.4 CPU years = 5.6 days using 1200 cores (with 60% efficiency)

$$F_{i+\frac{1}{2}} = \left(1 - \theta_{i+\frac{1}{2}}\right) F^{LF}_{i+\frac{1}{2}} + \theta_{i+\frac{1}{2}} F^{\text{WENO}}_{i+\frac{1}{2}} \qquad (8.169)$$

where $0 < \theta_{i+\frac{1}{2}} \leq 1$ is calculated by the positivity-preserving algorithm [21], with now

$$\alpha \frac{\Delta t}{\Delta x} < \frac{1}{2}. \tag{8.170}$$

Our astrophysical jets (see Fig. 8.21) WENO3 program is parallelized [15] via domain decomposition using MPI to distribute the computational grid over multiple processors.

8.24 WENO3 Simulations of Riemann Problems

In this section, we will compare a WENO3 simulation of the Toro/Sod Riemann problem

$$(\rho_l = 1, u_l = 0.75, P_l = 1) \text{ for } x < 0, \quad (\rho_r = 0.125, u_r = 0, P_r = 0.1) \text{ for } x > 0 \tag{8.171}$$

with $\gamma = 1.4$ (air) with the exact Riemann problem solution illustrated in Fig. 8.22. This Riemann problem models a laboratory shock-tube experiment, where a rubber membrane separating the two gases is ruptured with an electrical spark at $t = 0$. The left gas can be given a nonzero initial velocity by means of a moving (at $t = 0$) piston at the left end of the shock tube. The interaction of the two gases produces a leading shock wave with velocity 2.15, then a contact wave with velocity 1.36, and finally a trailing rarefaction wave with left boundary velocity -0.433 and right boundary velocity 0.300.

The type of each of the three waves in the Riemann problem solution is identified by using the Rankine-Hugoniot jump conditions. The left wave is a rarefaction, since density, pressure, and velocity (see Figs. 8.23, 8.24 and 8.25) vary smoothly through the wave, and the gas is rarefied through the wave; the middle wave is a contact since density jumps at the wave, but velocity and pressure are continuous; and the right wave is a shock since density, velocity, and pressure jump at the wave, and density and pressure are higher behind the wave.

Note that while WENO3 yields excellent resolution of the rarefaction wave and the shock wave, the contact wave is smeared out. This is typical of all shock/contact-capturing schemes, including ENO/WENO schemes, PPM, and CLAWPACK. To better resolve the contact, a contact tracking subgrid may be introduced (as in the front-tracking scheme of Glimm et al. [12]), or an additional level-set PDE for the contact may be introduced [32].

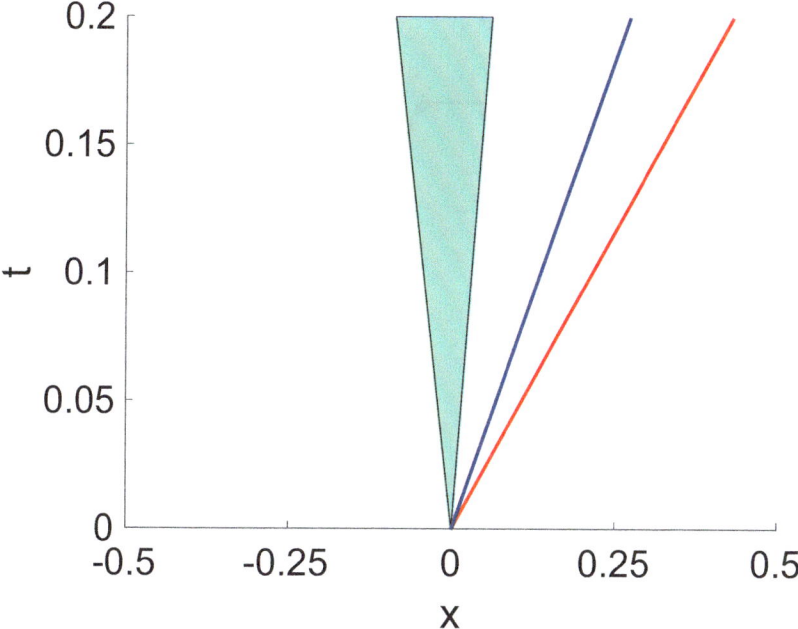

Fig. 8.22 Exact solution of the Toro/Sod Riemann problem in the *x-t* plane showing (from the left) a rarefaction wave (*shaded region*), a contact wave, and a shock wave

8.25 2D WENO3 Simulations of Supersonic Jets

In 2D, the Euler equations (8.106)–(8.108) of ideal gas dynamics take the following form (here $\mathbf{u} = (u, v)$):

$$\frac{\partial \rho}{\partial t} + \frac{\partial}{\partial x}(\rho u) + \frac{\partial}{\partial y}(\rho v) = 0 \tag{8.172}$$

$$\frac{\partial}{\partial t}(\rho u) + \frac{\partial}{\partial x}(\rho u^2 + P) + \frac{\partial}{\partial y}(\rho uv) = 0 \tag{8.173}$$

$$\frac{\partial}{\partial t}(\rho v) + \frac{\partial}{\partial x}(\rho uv) + \frac{\partial}{\partial y}(\rho v^2 + P) = 0 \tag{8.174}$$

$$\frac{\partial E}{\partial t} + \frac{\partial}{\partial x}(u(E+P)) + \frac{\partial}{\partial y}(v(E+P)) = 0. \tag{8.175}$$

These conservation laws are compactly expressed as

$$q_t + f(q)_x + g(q)_y = 0 \tag{8.176}$$

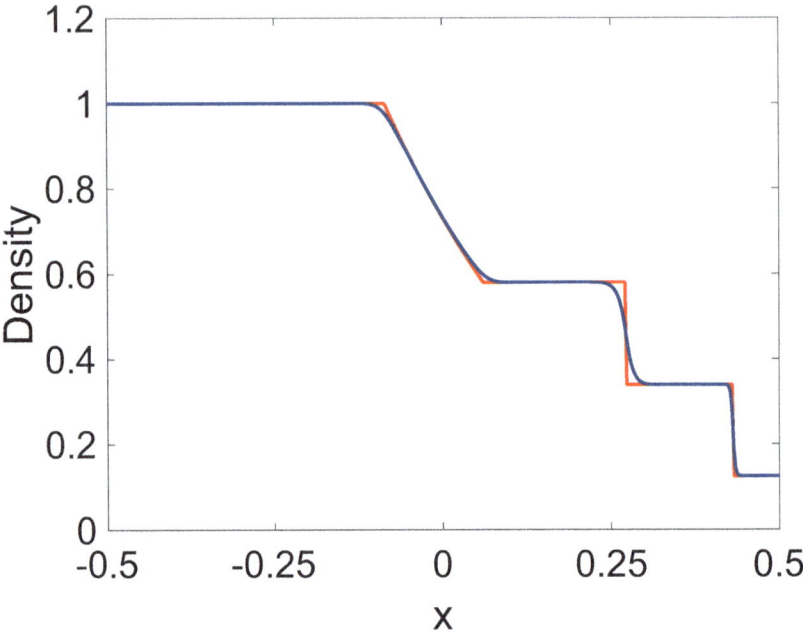

Fig. 8.23 Density for the Toro/Sod Riemann problem computed with WENO3 using **weno3.m** with $r = 0.9$ and $400\Delta x$ at $t = 0.2$ vs. exact (*sharper curve*) density, with (from the left) rarefaction wave, contact wave, and shock wave

where the conserved variables are

$$q = \begin{bmatrix} \rho \\ \rho u \\ \rho v \\ E \end{bmatrix} \qquad (8.177)$$

and the fluxes are

$$f(q) = \begin{bmatrix} \rho u \\ \rho u^2 + P \\ \rho u v \\ u(E+P) \end{bmatrix}, \quad g(q) = \begin{bmatrix} \rho v \\ \rho u v \\ \rho v^2 + P \\ v(E+P) \end{bmatrix}. \qquad (8.178)$$

In realistic applications to astrophysical jets, 2D cylindrically symmetric or fully 3D simulations should be performed, and there is a source term $-C(n, T)$ on the right-hand side of (8.175) from atomic and molecular radiative cooling. Here, however, we will consider the 2D Cartesian problem (8.172)–(8.175) without cooling for simplicity. Without the cooling term, the gas dynamical equations

8.25 2D WENO3 Simulations of Supersonic Jets

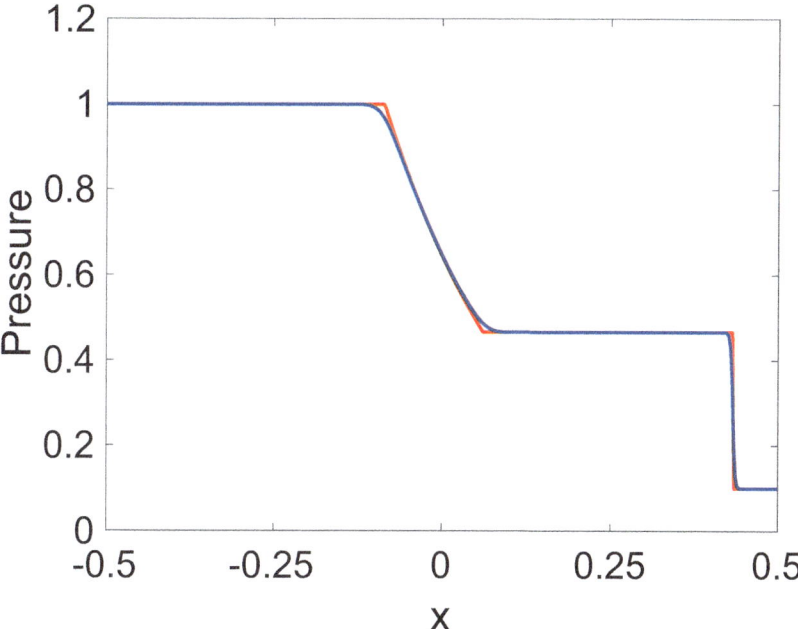

Fig. 8.24 Pressure for the Toro/Sod Riemann problem computed with WENO3 using **weno3.m** with $r = 0.9$ and $400\Delta x$ at $t = 0.2$ vs. exact (*sharper curve*) pressure

are invariant under a similarity transformation $x_i \to ax_i$, $t \to at$, so in this approximation micro-jets, aircraft/spacecraft jets, and (proto-stellar, stellar, and active galactic nuclei) astrophysical jets would all look the same, though magnified by the factor a.

Boundary conditions for the jet simulations are through-flow, except for the jet inflow boundary condition at the middle of the left boundary, where $\rho = \rho_{\text{jet}}$, $u = u_{\text{jet}}$, $v = 0$, and $T = T_{\text{jet}}$ are specified. The initial conditions set the ambient values ρ_{amb}, $u_{\text{amb}} = 0$, $v_{\text{amb}} = 0$, and T_{amb}. Recall that for the ideal gas $P = nk_BT$ with $n = \rho/m$, and $E = \frac{3}{2}nk_BT + \frac{1}{2}\rho(u^2 + v^2)$ for the monatomic gas of H atoms with $\gamma = 5/3$.

Due to the dimension-by-dimension splitting employed in the WENO method, the 2D code simply involves two copies of the 1D code, one for the x sweep and one for the y sweep.

2D Cartesian supersonic jet simulations are presented in Figs. 8.26 and 8.27, using the WENO3 **jets.m** code: a Mach 80 heavy pressure-matched jet and a Mach 6 heavy jet with $P_{\text{jet}} = P_{\text{amb}}/5$ showing shock diamonds, respectively. Here the Mach numbers are for the jet inflow velocity with respect to the jet gas. Compare the Mach 80 jets in Figs. 1.1 and 8.26: The jet in the first figure includes radiative cooling, which suppresses the Kelvin-Helmholtz rollup of the jet tip.

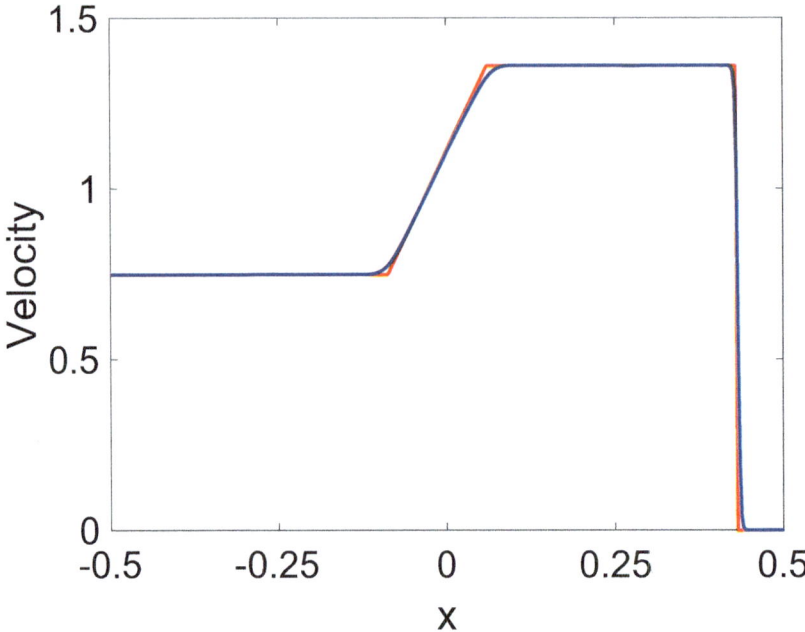

Fig. 8.25 Velocity for the Toro/Sod Riemann problem computed with WENO3 using **weno3.m** with $r = 0.9$ and $400\Delta x$ at $t = 0.2$ vs. exact (*sharper curve*) velocity

Figure 8.26 illustrates a bow shock in front of the jet, Kelvin-Helmholtz rollup of the tip of the jet, and a terminal shock system (Mach disk) inside the jet near the tip—which brings the inflow jet velocity down to match the propagation velocity of the tip of the jet.

The jet in Fig. 8.27 exhibits similar features, except that the bow shock is now a bow wave, and a crisscross pattern of shock diamonds (X-shaped interacting shocks) develops inside the jet stem.

8.26 WENO3 Codes

The WENO3 method is employed in five of our numerical codes:

burgers3.m for the inviscid Burgers equation
wave3.m for the linear wave equation
weno3.m for gas dynamics
jets.m for 2D supersonic jets
chd.m for the classical hydrodynamic model.

8.26 WENO3 Codes

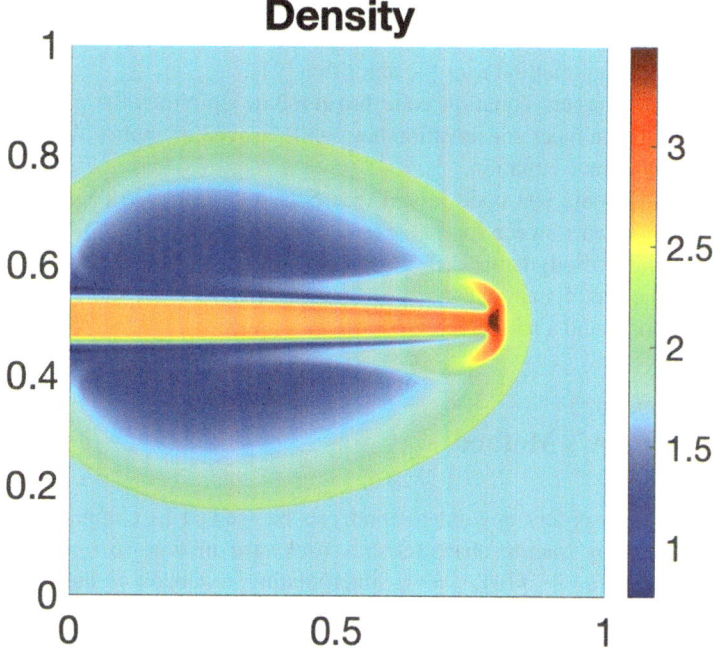

Fig. 8.26 WENO3 simulation of $\log_{10}(n)$ (with n in H/cm^3) using **jets.m** of a 2D Mach 80 heavy supersonic jet at $t = 11$ year on a $500\Delta x \times 500\Delta y$ grid with $r = 0.25$. The grid is 10^{11} km on each side

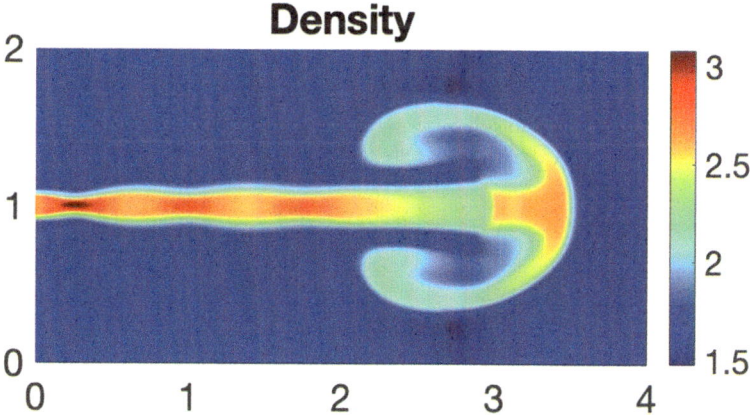

Fig. 8.27 WENO3 simulation of $\log_{10}(n)$ (with n in H/cm^3) using **jets.m** of a 2D Mach 6 heavy supersonic jet at $t = 1750$ year on a $400\Delta x \times 200\Delta y$ grid with $r = 0.9$ and $P_{jet} = P_{amb}/5$, displaying shock diamonds. The grid is $(4, 2) \times 10^{11}$ km

Our WENO3 code was developed first in Fortran 90 for astrophysical jet simulations and then ported for pedagogical purposes to MATLAB by Jeremiah Jones and the author, based on a Fortran code of Guang-Shan Jiang and Chi-Wang Shu.

The code in **weno3.m** for transforming to characteristics variables uses the left and right Roe matrices [30] before calculating the WENO3 fluxes; the code for the positivity-preserving method uses the algorithm [21].

The inviscid Burgers equation code **burgers3.m** simplifies the gas dynamical code to a scalar nonlinear conservation law, while **wave3.m** applies WENO3 to the linear first-order wave equation.

The 2D supersonic jets code **jets.m** (see Sect. 8.25) has two copies of the 1D WENO3 method: an x sweep for $q_t + f(q)_x = 0$ and a y sweep for $q_t + g(q)_y = 0$.

The classical hydrodynamic code **chd.m** (see Sect. 9.5) implements the WENO3 method for the gas dynamical transport modes, TRBDF2 for the parabolic heat conduction mode, and a tridiagonal direct solve for the elliptic Poisson equation.

8.27 Godunov's Method

The original (first-order) upwind method can be viewed as computing the new solution at t_{n+1} by tracing characteristics backward in time to t_n. For a scalar conservation law $u_t + f(u)_x = 0$, interpolating to obtain u the foot of the characteristic flowing into (x_i, t_{n+1}) produces the nonconservative upwind scheme:

$$u_i^{n+1} = \frac{1}{\Delta x}\left[(\Delta x - f'(u_i^n)\Delta t)\, u_i^n + f'(u_i^n)\Delta t\, u_{i-1}^n\right]$$

$$= u_i^n - \frac{\Delta t}{\Delta x} f'(u_i^n)(u_i^n - u_{i-1}^n) \tag{8.179}$$

for $f'(u_i^n) > 0$, and

$$u_i^{n+1} = \frac{1}{\Delta x}\left[(\Delta x + f'(u_i^n)\Delta t)\, u_i^n - f'(u_i^n)\Delta t\, u_{i+1}^n\right]$$

$$= u_i^n - \frac{\Delta t}{\Delta x} f'(u_i^n)(u_{i+1}^n - u_i^n) \tag{8.180}$$

for $f'(u_i^n) < 0$.

Godunov's method (1959) [17] revolutionized the development of hyperbolic methods for conservation laws by advocating that solution updates be calculated by (approximately) solving Riemann problems forward in time rather than by (approximately) tracing characteristics backward in time. Godunov's method is conservative, since Riemann problems provide exact solutions to the conservation laws.

Define a piecewise constant function $\widetilde{w}(x, t_n) = w_i^n$ for grid cell i with $x_{i-\frac{1}{2}} < x < x_{i+\frac{1}{2}}$ (this makes the method first-order accurate). Godunov's method computes $\widetilde{w}(x, t)$ for $t_n \leq t \leq t_{n+1}$ from Riemann problem solutions (see Fig. 8.28).

8.27 Godunov's Method

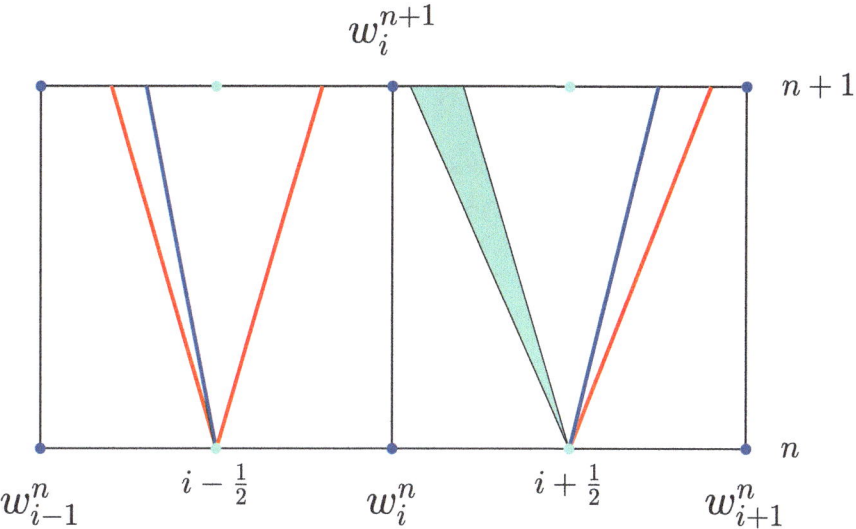

Fig. 8.28 Space-time stencil for Godunov's method. Grid cell i runs from $x_{i-\frac{1}{2}}$ to $x_{i+\frac{1}{2}}$. Piecewise constant values w_i^n and w_{i+1}^n over grid cells i and $i + 1$, respectively, define a Riemann problem at grid point $i + \frac{1}{2}$, etc.

First assume that Δt is sufficiently small ($\Delta t \leq \frac{1}{2}\Delta x / \max\{|\lambda|\}$, where $\max\{|\lambda|\}$ is taken over all characteristic speeds and all grid points at t_n) that the Riemann problem solutions at the $x_{i+\frac{1}{2}}$ do not interact for all i. Define w_i^{n+1} as a cell average:

$$w_i^{n+1} \equiv \frac{1}{\Delta x} \int_{x_{i-\frac{1}{2}}}^{x_{i+\frac{1}{2}}} \widetilde{w}(x, t_{n+1}) \, dx. \tag{8.181}$$

Then integrating the conservation law $w_t + f(w)_x = 0$ over a space-time rectangle $x_{i-\frac{1}{2}} \leq x \leq x_{i+\frac{1}{2}}$ and $t_n \leq t \leq t_{n+1}$, we obtain a conservative space-time finite volume method

$$w_i^{n+1} = \frac{1}{\Delta x} \left[\int_{x_{i-\frac{1}{2}}}^{x_{i+\frac{1}{2}}} \widetilde{w}(x, t_n) \, dx - \int_{t_n}^{t_{n+1}} f\left(\widetilde{w}\left(x_{i+\frac{1}{2}}, t\right)\right) dt \right.$$
$$\left. + \int_{t_n}^{t_{n+1}} f\left(\widetilde{w}\left(x_{i-\frac{1}{2}}, t\right)\right) dt \right] \tag{8.182}$$

or

$$w_i^{n+1} = w_i^n - \frac{\Delta t}{\Delta x} \left[F(w_i^n, w_{i+1}^n) - F(w_{i-1}^n, w_i^n) \right] \tag{8.183}$$

where

$$F(w_i^n, w_{i+1}^n) = \frac{1}{\Delta t} \int_{t_n}^{t_{n+1}} f\left(\widetilde{w}\left(x_{i+\frac{1}{2}}, t\right)\right) dt = f(w^*(w_i^n, w_{i+1}^n)). \qquad (8.184)$$

Here we have made use of the fact that since the solution to the Riemann problem is self-similar, $\widetilde{w}\left(x_{i+\frac{1}{2}}, t\right) = w^*$ is constant. The CFL condition $\Delta t \leq \Delta x / \max\{|\lambda|\}$ insures that even if the neighboring Riemann problems interact, the interaction will not affect $\widetilde{w}\left(x_{i+\frac{1}{2}}, t\right) = const$.

Thus, Godunov's method can be elegantly written in the conservative form

$$w_i^{n+1} = w_i^n - \frac{\Delta t}{\Delta x} \left[f(w^*(w_i^n, w_{i+1}^n)) - f(w^*(w_{i-1}^n, w_i^n)) \right]. \qquad (8.185)$$

Higher-order Godunov methods like CLAWPACK or PPM construct $\widetilde{w}(x, t_n)$ to be piecewise linear or piecewise parabolic. All the Godunov methods use a finite volume *Reconstruct–Evolve–Average* algorithm (see LeVeque [24, 25]):

- *Reconstruct* $\widetilde{w}(x, t_n)$ from w_i^n as piecewise constant, piecewise linear, piecewise parabolic, etc.
- *Evolve* Riemann problems exactly[7] or approximately with this initial data to obtain $\widetilde{w}(x, t_{n+1})$.
- *Average* to get

$$w_i^{n+1} \equiv \frac{1}{\Delta x} \int_{x_{i-\frac{1}{2}}}^{x_{i+\frac{1}{2}}} \widetilde{w}(x, t_{n+1}) \, dx. \qquad (8.186)$$

CLAWPACK is piecewise linear and second-order accurate in smooth regions;[8]; PPM is third-order accurate in smooth regions. Higher-order Godunov methods are conservative as long as the *Reconstruct* step is conservative (as it is for Godunov's method, PPM, and CLAWPACK), since the *Evolve* (Riemann problem solutions) and *Average* (cell averaging) steps are always conservative.

Define the total variation of a solution as

$$\text{TV}(w) = \sum_i |w_i - w_{i-1}|. \qquad (8.187)$$

Oscillations in a numerical solution (e.g., the Gibbs overshoot around a shock wave) cause the total variation to increase in time. A desired quality for a hyperbolic

---

[7] For gas dynamics, the Riemann problems can be solved exactly for γ-law gases, but only with piecewise constant initial data.
[8] There is loss of accuracy (typically to first order) for all shock-capturing schemes at discontinuities.

numerical method is that it be total variation diminishing (TVD): $TV(w^{n+1}) \leq TV(w^n)$. Both PPM and CLAWPACK are TVD. Some versions of ENO and WENO are TVD, but the WENO3-LF method described here is not, and there may be small oscillations around a strong shock wave.

8.28 Loss of Accuracy at a Shock Wave

The error function is defined by

$$\text{erf}(x) = \frac{2}{\sqrt{\pi}} \int_0^x e^{-y^2} dy. \tag{8.188}$$

The complementary error function is defined by

$$\text{erfc}(x) = 1 - \text{erf}(x) = \frac{2}{\sqrt{\pi}} \int_x^\infty e^{-y^2} dy. \tag{8.189}$$

Note that erf(x) goes from -1 to 1 as x goes from $-\infty$ to ∞, while erfc(x) goes from 2 to 0 as x goes from $-\infty$ to ∞ (see Fig. 8.29).

For a first-order hyperbolic method like Lax-Friedrichs or upwind, the modified PDE for the wave equation $u_t + cu_x = 0$, $c > 0$, is the advection-diffusion equation $u_t + cu_x = Du_{xx}$, with

$$D_{LF} = \frac{c\Delta x}{2r}(1-r^2), \quad D_{\text{upwind}} = \frac{c\Delta x}{2}(1-r) \tag{8.190}$$

where the Courant number $r = c\Delta t/\Delta x \leq 1$.

We first derive the shock profile solution for the advection-diffusion equation

$$u_t + cu_x = Du_{xx}, \quad u_0(x) = \begin{cases} 2 & x \leq 0 \\ 0 & x > 0 \end{cases} \tag{8.191}$$

Setting $u(x, t) = w(x - at, t)$, we obtain

$$w_t = Dw_{xx}, \quad w_0(x) = \begin{cases} 2 & x \leq 0 \\ 0 & x > 0 \end{cases} \tag{8.192}$$

The general solution of the diffusion equation (8.192) for shock wave initial data is

$$w_D(x, t) = \frac{1}{\sqrt{4\pi Dt}} \int_{-\infty}^\infty \exp\left\{-\frac{(x-y)^2}{4Dt}\right\} w_0(y)\, dy$$

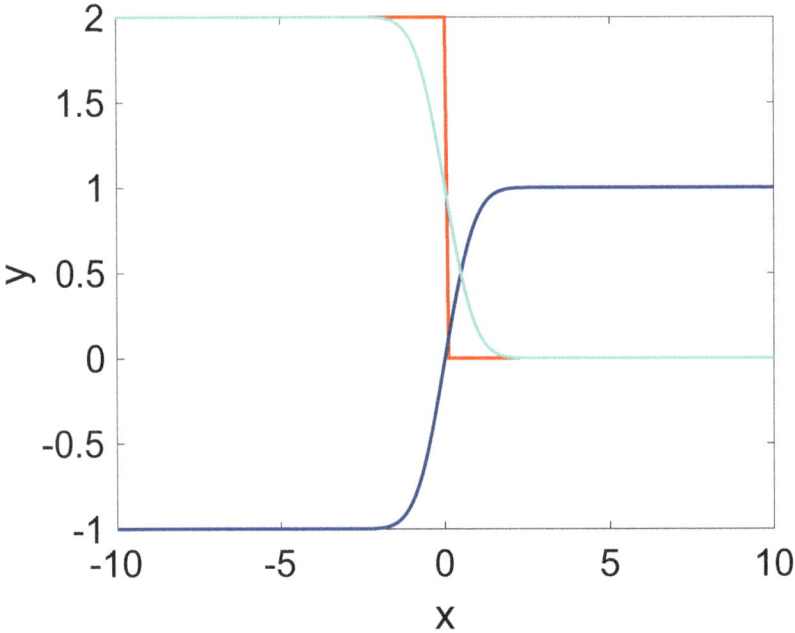

Fig. 8.29 Error functions erf (*dark smooth curve*) and erfc (*light smooth curve*), plus shock wave (*discontinuous curve*) solution

$$= \frac{1}{\sqrt{4\pi Dt}} \int_{-\infty}^{\infty} \exp\left\{-\frac{y^2}{4Dt}\right\} w_0(x-y)\, dy$$

$$= \frac{1}{\sqrt{\pi Dt}} \int_{x}^{\infty} \exp\left\{-\frac{y^2}{4Dt}\right\} dy = \operatorname{erfc}\left\{\frac{x}{\sqrt{4Dt}}\right\}. \quad (8.193)$$

Now we can calculate the norm of the difference between the shock wave exact solution ($D = 0$) and the shock profile diffusive solution:

$$\|u(\cdot, t) - u_D(\cdot, t)\|_1 = \|w(\cdot, t) - w_D(\cdot, t)\|_1$$

$$= 2 \int_0^{\infty} \operatorname{erfc}\left\{\frac{x}{\sqrt{4Dt}}\right\} dx = \frac{4\sqrt{Dt}}{\sqrt{\pi}} \sim \sqrt{t\, \Delta x} \quad (8.194)$$

as $\Delta x \to 0$ with r fixed. Thus the first-order methods become only $\frac{1}{2}$-order accurate in the presence of a shock wave. Heuristically, the numerical methods are first-order accurate in smooth regions of flow and lose accuracy mainly near the discontinuity. High-order schemes for gas dynamics are high order in smooth regions of flow and typically degenerate to first order near discontinuities.

Problems

8.1 Derive the modified PDE for the Lax-Friedrichs method[9] for $u_t + cu_x = 0$. Find the coefficient $D_n \sim \{\Delta t, \Delta x\}$ of numerical diffusion in $u_t + cu_x = D_n u_{xx}$. Show $D_n \geq 0$ if the Courant number $r = c\Delta t/\Delta x \leq 1$. (Since $D_n u_{xx} = -\tau$, this shows Lax-Friedrichs is first order. Note that spatial and temporal errors are intertwined in D_n through the factors of r.)

8.2 Show the Lax-Friedrichs method is conditionally stable for $u_t + cu_x = 0$ using von Neumann stability analysis. *Hint:* Show $|G(k)|^2 = G^*(k)G(k) \leq 1$ if $r \leq 1$.

8.3 Show all four versions (8.105) of Lax-Friedrichs are conservative by verifying that the numerical flux function

$$F_{i+\frac{1}{2}} = \frac{1}{2}(f(w_i) + f(w_{i+1})) - \frac{\Delta x}{2\Delta t}\left(r_{i+\frac{1}{2}}w_{i+1} - r_{i+\frac{1}{2}}w_i\right)$$

correctly produces the various Lax-Friedrichs methods for $w_t + f(w)_x = 0$.

8.4 Using von Neumann stability analysis, show downwind is unconditionally unstable for $u_t + cu_x = 0$. *Hint:* Show $|G(k)|^2 = G^*(k)G(k) > 1$ for any value of $r > 0$.

8.5 Using von Neumann stability analysis, show Lax-Wendroff is stable for $u_t + cu_x = 0$ as long as the CFL condition $r \leq 1$ is satisfied.

8.6 In the modified PDE for the Lax-Wendroff method for $u_t + cu_x = 0$, derive the coefficient $\beta = \frac{ch^2}{6}(r^2 - 1)$ of numerical dispersion in $u_t + cu_x = \beta u_{xxx}$. (Incidentally, this shows Lax-Wendroff is second-order accurate.)

8.7 Show the two-step Lax-Wendroff method reduces to the original Lax-Wendroff scheme for $u_t + Au_x = 0$, where A is a constant matrix.

8.8 Confirm that the Lax-Friedrichs and Lax-Wendroff methods produce the exact solution projected onto the grid for $r = 1$ for the first-order and second-order wave equations. *Hint:* See **wave2.m**.

8.9 Show the advection-diffusion equation $u_t + cu_x = \nu u_{xx}$ has the vanishing viscosity solution $u_0(x - ct)$ as $\nu \to 0$. *Hints:* First show $w(x, t) = u(x + ct, t)$ satisfies the diffusion equation $w_t = \nu w_{xx}$, and then take $\nu \to 0$ in the general solution to the diffusion equation. Finally use the fact that $u(x, t) = w(x - ct, t)$. The general solution to the diffusion equation $w_t = \nu w_{xx}$ is

$$w(x, t) = \int_{-\infty}^{\infty} K(x - y, t) w_0(y)\, dy = \frac{1}{\sqrt{4\pi \nu t}} \int_{-\infty}^{\infty} \exp\left\{-\frac{(x - y)^2}{4\nu t}\right\} w_0(y)\, dy$$

---

[9] All four versions of Lax-Friedrichs are CFL stable, first-order accurate, diffusive, and conservative, but it is sufficient just to analyze the original method in this problem and the next.

with

$$\lim_{t \to 0} K(x,t) = \lim_{\nu \to 0} K(x,t) = \delta(x).$$

8.10

(a) Show Burgers' equation $u_t + uu_x = \nu u_{xx}$ has a viscous shock profile traveling wave solution of the form $u(x,t) = w(x - st)$ by deriving an ODE for w and showing the ODE is solved by

$$w(y) = u_r + \frac{1}{2}(u_l - u_r)\left[1 - \tanh\frac{(u_l - u_r)y}{4\nu}\right], \quad u_l > u_r$$

with $s = (u_l + u_r)/2$.

(b) Show $u(x \to -\infty, t) = u_l$ and $u(x \to \infty, t) = u_r$.

(c) Show the vanishing viscosity limit $\nu \to 0$ of $w(x - st)$ gives the shock wave solution to the Riemann problem $(u_l > u_r)$ for the inviscid Burgers equation:

$$u(x,t) = \begin{cases} u_l & x < st \\ u_r & x > st \end{cases} = u_0(x - st)$$

with shock velocity $s = (u_l + u_r)/2$.

8.11 Show that for the inviscid Burgers equation $u_t + uu_x = 0$ with smooth initial data $u_0(x)$ and $u_0'(x) < 0$ at some points, the breaking time for shock formation is

$$t_b = \frac{-1}{\min\{u_0'(x)\}}.$$

Hint: You can use the Implicit Function Theorem or the geometry of characteristics in the x-t plane.

8.12 Prove the Lax-Friedrichs method is positivity-preserving (i.e., if $u_i^n > 0$ for all i, then $u_i^{n+1} > 0$ for all i) for the inviscid Burgers equation $u_t + \left(\frac{1}{2}u^2\right)_x = 0$. The LF discretization of the inviscid Burgers equation is

$$u_i^{n+1} = \frac{1}{2}(u_{i-1}^n + u_{i+1}^n) - \frac{\Delta t}{4\Delta x}((u_{i+1}^n)^2 - (u_{i-1}^n)^2).$$

Hint: For stability, $\Delta t \leq \Delta x/\max_i\{|u_i^n|\}$. *Note:* Lax-Friedrichs is also positivity-preserving for gas dynamics, unlike upwind schemes.

8.13 Show the upwind discretization of the inviscid Burgers equation $u_t + uu_x = 0$ for $u(x,t) > 0$

Problems

$$u_i^{n+1} = u_i^n - \frac{\Delta t}{\Delta x} u_i^n (u_i^n - u_{i-1}^n)$$

is nonconservative, while the upwind discretization of the inviscid Burgers equation $u_t + \left(\frac{1}{2}u^2\right)_x = 0$ for $u(x,t) > 0$

$$u_i^{n+1} = u_i^n - \frac{\Delta t}{2\Delta x} \left((u_i^n)^2 - (u_{i-1}^n)^2\right)$$

is conservative by showing whether or not the scheme can be put into the form

$$u_i^{n+1} = u_i^n - \frac{\Delta t}{\Delta x} \left(F_{i+\frac{1}{2}}^n - F_{i-\frac{1}{2}}^n\right).$$

8.14 Using **burgers1.m** and **burgers3.m** with $200\Delta x$ and $r = 0.9$, solve the four problem types (Riemann problem shock, Riemann problem rarefaction, merging shocks, N-wave) using six different numerical methods: conservative upwind, the four different versions of Lax-Friedrichs, and WENO3. Discuss your results. Which are the best methods for the different problems?

8.15 For the inviscid Burgers equation, show conservative upwind, (the best version of) Lax-Friedrichs, and WENO3 converge under mesh refinement by comparing and contrasting simulations of the Riemann problem shock and the Riemann problem rarefaction with $r = 0.9$ and $N = 50, 100,$ and $200\Delta x$. Use

```
burgers1(200,120,0.9,1,1) % for Riemann problem shock
burgers1(200,100,0.9,1,2) % for Riemann problem rarefaction
```

etc. *Hint:* Since $\Delta t \sim \Delta x$ for hyperbolic PDEs, if you double the number N of Δx, you also have to double the number of timesteps to have the same final time.

8.16

(a) Derive the 1D entropy advection equation $s_t + us_x = 0$ (valid only away from discontinuities) from the 1D Euler equations and the expression for the entropy $s = c_V \ln(P/\rho^\gamma)$ of a polytropic gas. *Hint:* Start with the entropy advection equation, and then derive the energy conservation equation $E_t + (u(E+P))_x = 0$ with $P = (\gamma - 1)\left(E - \frac{1}{2}\rho u^2\right)$.
(b) Show the entropy density $S = \rho s/m$ satisfies the entropy conservation law $S_t + (uS)_x = 0$, away from discontinuities.

8.17 For the 1D Euler equations $w_t + f(w)_x = 0$, derive the Jacobian matrix $f'(w) = \partial f/\partial w$. Then show:

(a) $\text{Tr}\{f'(w)\} = 3u = \lambda_0 + \lambda_+ + \lambda_-$, where $\lambda_0 = u$ and $\lambda_\pm = u \pm c$,
(b) $\det\{f'(w)\} = u(u^2 - c^2) = \lambda_0 \lambda_+ \lambda_-$ where $c^2 = \gamma P/\rho$, and
(c) the eigenvector $r_2 = (1, u, u^2/2)$ for the eigenvalue λ_0.

8.18 Switch to physical variables $w = [n, u, T]$ for 1D gas dynamics, and derive equations for w_t. Then show the eigenvalues σ of the symbol S of the linearized PDE system are $\sigma_0 = iku$ and $\sigma_\pm = ik(u \pm c)$ (all hyperbolic modes). Here σ is the decay rate of the perturbation $e^{-\sigma t + ikx} \delta w$. *Hint:* You can simply verify the Tr$\{S\}$ and det$\{S\}$ formulas for the eigenvalues.

8.19

(a) Show the solution of the Riemann problem

$$\rho_l = 2, u_l = 1, P_l = 3; \quad \rho_r = 1, u_r = 0, P_r = 1$$

with $\gamma = 1.5$ is a single shock wave propagating to the right. Calculate the shock velocity s. For $\gamma = 1.5$, $E = \frac{1}{2}\rho u^2 + 2P$. *Hint:* Use the jump conditions.

(b) With a wall boundary condition at $x = 1$, analytically calculate the solution after reflection of the shock wave.

8.20 Derive equation (8.119) from the Rankine-Hugoniot jump conditions (8.113)–(8.117). Then derive the relation governing the pressure jump P_b/P_a (8.120) across the shock. Show the limiting strong shock density jump is obtained by taking $P_b/P_a \to \infty$: $\rho_b/\rho_a = 1/\mu^2$. *Note:* For a monatomic gas like H with $\gamma = 5/3$, the limiting strong shock density jump is $\rho_b/\rho_a = 4$, while for air with $\gamma = 7/5$, the limiting $\rho_b/\rho_a = 6$. Much stronger density contrasts are attained in real-world situations like astrophysical shocks with radiative cooling, where the density ratio can be arbitrarily high.

8.21 Complete the proof that the entropy s_b behind a shock is greater than the entropy s_a ahead of the shock by showing in (8.121) that

$$\varphi(X) = \frac{X - \mu^2}{(1 - \mu^2 X) X^\gamma}, \quad 1 < X < 1/\mu^2$$

increases monotonically from $\varphi(1) = 1$ (and thus that $\ln(\varphi(X)) > 0$), where $X = \rho_b/\rho_a$, with $1 < \gamma$ and $0 < \mu^2 < 1$. *Hint:* Show

$$\frac{d\varphi}{dX} = \frac{\gamma \mu^2 (X - 1)^2}{(1 - \mu^2 X)^2 X^{\gamma+1}}$$

which is positive for $1 < X < 1/\mu^2$.

8.22 Using **weno3.m**, investigate the effects of the Courant number r on the solution of the Toro/Sod Riemann problem

$$\rho_l = 1, u_l = 0.75, p_l = 1; \quad \rho_r = 0.125, u_r = 0, p_r = 0.1$$

with $\gamma = 1.4$. Take $200\Delta x$ and Courant number $r = 0.1, 0.5, 0.9$. Use

Problems 181

```
weno3(1,0.75,1,0.125,0,0.1,1.4,0.2,200,0.1,0)
```

etc. Compare the Density plots (computed vs. exact solution) at time $t = 0.2$ for each of three cases. Briefly discuss your results.

8.23 Show **weno3.m** converges under mesh refinement by simulating the Toro/Sod Riemann problem with 100, 200, and 400Δx, with $r = 0.9$. Use

```
weno3(1,0.75,1,0.125,0,0.1,1.4,0.2,100,0.9,0)
```

etc. Compare the Density plots (computed vs. exact solution) at time $t = 0.2$ for each of three cases. Discuss your results.

8.24 Compare WENO3 with and without transforming to characteristic variables on the Riemann problems in **weno3.m**.

8.25 Show **jets.m** converges under mesh refinement by simulating the heavy Mach 80 supersonic jet with 100$\Delta x \times$ 50Δy, 200$\Delta x \times$ 100Δy, and 400$\Delta x \times$ 200Δy, with $r = 0.25$. Use

```
jets(500,300,0,10^3,50,0,0,10^4,5/3,1,1,100,100,11,0.25)
```

etc. Compare the Density plots at time $t = 260$ for each of three cases. Discuss your results.

8.26 In the WENO3 method (for simplicity, assume the flow is to the right), suppose $\beta_1^+ = (f_i - f_{i-1})^2 = 0$ (smooth flow between grid points $i - 1$ and i).

(a) If in addition $\beta_2^+ = (f_{i+1} - f_i)^2 = 0$, show $\omega_1^+ = \gamma_1$ and $\omega_2^+ = \gamma_2$, which gives the linear scheme for $F_{i+\frac{1}{2}}$.

(b) If instead $\beta_2^+/\epsilon \to \infty$, show $\omega_1^+ = 1$ and $\omega_2^+ = 0$, which reproduces the original ENO scheme for $F_{i+\frac{1}{2}}$.

8.27 Verify that the third-order WENO method is in fact third order in regions of smooth flow (assume the flow is to the right) by showing

$$\frac{F_{i+\frac{1}{2}} - F_{i-\frac{1}{2}}}{\Delta x} = f'_i + \frac{\Delta x^3}{12} f_i^{(4)} + \cdots$$

Hint: First verify the error formulas for $F^{(1)}_{i\pm\frac{1}{2}}$ and $F^{(2)}_{i\pm\frac{1}{2}}$, and then use the linear weights in calculating $F_{i\pm\frac{1}{2}}$.

8.28 Verify RK3 (8.164)–(8.166) is third-order in time for the linear problem $\mathcal{L}u = au$. *Hint:* Show $u^{n+1} = Gu^n$ agrees with the exact $e^{a\Delta t}u^n$ through third-order in Δt.

8.29 *Project:* Add the positivity-preserving method from **weno3.m** to **jets.m**. How far can the new code go in Mach number for the 2D heavy supersonic jet in Fig. 8.26, varying just ujet in

```
jets(500,ujet,0,10^3,50,0,0,10^4,5/3,1,1,500,500,11,0.25)
```

Hint: Start with numerical experiments on a $100\Delta x \times 100\Delta y$ grid.

8.30 *Project:* It is straightforward to show our upwind, Lax-Friedrichs, and Lax-Wendroff codes obtain the correct orders of accuracy for the wave equations when the Courant number $r = 0.99$ (say), though larger diffusive or dispersive errors enter for $r \leq 0.95$. (Recall that these schemes give the exact solution for the wave equations projected onto the grid for $r = 1$.)

The situation is more complicated for nonlinear hyperbolic PDEs.

(a) Show that for the first-order wave equation with $u_0(x) = \sin^4(10\pi x)$ initial data that WENO3 is in fact third-order in regions of smooth flow. Use 1000, 2000, and $4000\Delta x$ in **wave3.m** with $t = 0.5$. See Fig. 8.30.

(b) If $u_0(x) = \sin(10\pi x)$, however, show WENO3 in **wave3.m** only obtains second-order accuracy.

(c) For the inviscid Burgers equation, show the conservative upwind method in **burgers1.m** is first-order accurate for the shock and rarefaction Riemann problems.

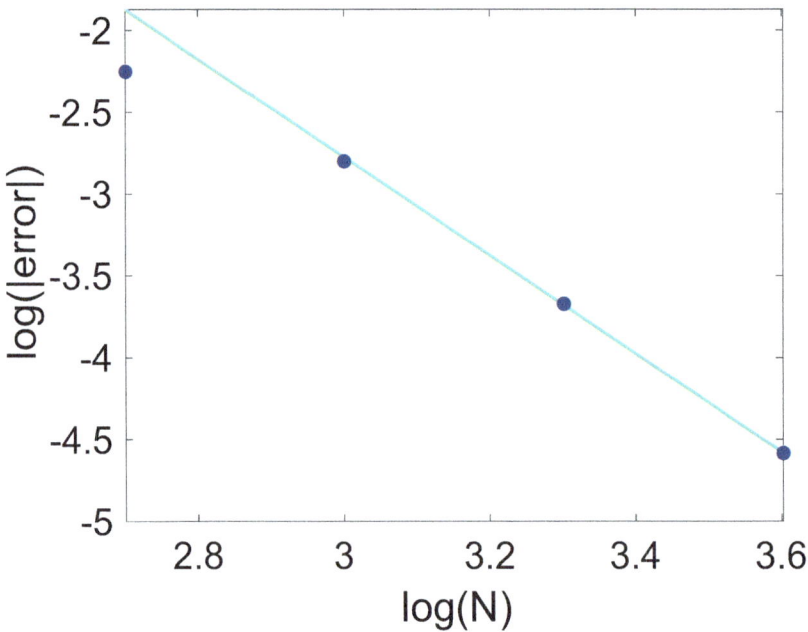

Fig. 8.30 Numerical errors $\log_{10}(||u_{\text{exact}} - u_{\text{comp}}||_1)$ vs. $\log_{10}(N)$ for WENO3 for the first-order wave equation using **wave3.m** with $r = 1$ and $N = 500, 1000, 2000$, and $4000\Delta x$ at $t = 0.5$. The *straight line* is the theoretical value with log-log slope of -3. Note that lower accuracy is obtained with $N = 500$ due to excessive numerical diffusion

(d) For the inviscid Burgers equation, show only second-order accuracy is obtained by WENO3 in **burgers3.m** for straight line initial data $u_0(x) = x - \frac{1}{2}$ and only first-order accuracy for the shock and rarefaction Riemann problems.
(e) For 1D gas dynamics, show WENO3 only obtains first-order accuracy for the eight Riemann problems in **weno3.m**.

These results are consistent with results in the literature: higher-order upwind methods are generally high order in regions of smooth flow but degenerate to first order near discontinuities and even near sharp gradients. Note that WENO5 performs much better in achieving the desired order of accuracy.

Also note that even though the computed solution to the gas dynamical Riemann problem in Fig. 8.23 is formally only first-order accurate, qualitatively it is an excellent fit to the exact solution (except for the smearing of the contact), with correct wave structure and correct wave speeds.

8.31 *Project:* Implement the conservative first-order upwind method and (the best) Lax-Friedrichs method into **weno3.m**, and then compare with each other and WENO3 on the Riemann problems in **weno3.m**.

8.32 *Project:* WENO5 gives much better spatial resolution than WENO3. Program the WENO5 method by modifying **weno3.m**, and compare WENO5 against WENO3 on the Riemann problems in **weno3.m**.

Chapter 9
Numerical Methods for Mixed Type PDEs

There are many PDE systems that contain hyperbolic, parabolic, and elliptic modes. For example, in the classical hydrodynamic model for semiconductor devices, the hyperbolic compressible fluid dynamics of electrons is coupled to parabolic heat conduction and/or viscosity and to the elliptic Poisson equation for the electric potential. The most general approach is to use fractional timesteps: A timestep can be split into a hyperbolic transport partial timestep, followed by a parabolic heat conduction/viscosity partial timestep, with an elliptic electric potential update at the end.

Detailed examples of timestep splittings will be given for the drift-diffusion model (parabolic plus elliptic), the incompressible Navier-Stokes equations (incompletely parabolic), and the classical hydrodynamic model (hyperbolic plus parabolic plus elliptic).

9.1 Strang Splitting

Timestep splitting can introduce errors from splitting a spatial operator term and a source term or from splitting two or more spatial operator terms.

Source terms are generally treated by a separate partial timestep. For example, in the linear advection-reaction equation

$$u_t = -cu_x + \alpha(x)u \equiv (A + B)u \tag{9.1}$$

typically we would update (using here for the sake of analysis the exact time evolution operator for each partial step)

$$u^* = e^{\Delta t A} u^n \tag{9.2}$$

$$u^{n+1} = e^{\Delta t B} u^* \tag{9.3}$$

while the exact solution is formally

$$u(t_{n+1}) = e^{\Delta t (A+B)} u^n = \sum_{j=0}^{\infty} (A+B)^j u^n. \tag{9.4}$$

The splitting error is

$$e_l = u(t_{n+1}) - u^{n+1} = \left(e^{\Delta t (A+B)} - e^{\Delta t B} e^{\Delta t A} \right) u^n$$
$$= \frac{1}{2} \Delta t^2 (AB - BA) u^n + O(\Delta t^3). \tag{9.5}$$

Thus, source term splitting introduces locally second-order and globally first-order errors if the commutator of A with B

$$AB - BA \equiv [A, B] \neq 0 \tag{9.6}$$

independent of the spatial and temporal orders of accuracy of the methods for the A and B partial steps. (In our example of advection-reaction, $[A, B]u = -c\alpha_x u$.)

Second-order accuracy can be achieved by Strang splitting: Approximate

$$e^{\Delta t (A+B)} u^n \approx e^{\frac{1}{2}\Delta t A} e^{\Delta t B} e^{\frac{1}{2}\Delta t A} u^n. \tag{9.7}$$

In other words, to advance from t_0 to t_n, first take a half timestep $\frac{1}{2}\Delta t$ with A, then take a full timestep with B, then a full timestep with A, ..., then a full timestep with B, then finally a half timestep $\frac{1}{2}\Delta t$ with A:

$$u^n \approx e^{\frac{1}{2}\Delta t A} e^{\Delta t B} e^{\Delta t A} \cdots e^{\Delta t B} e^{\frac{1}{2}\Delta t A} u_0. \tag{9.8}$$

Now the splitting error is $O(\Delta t^2)$ globally (see Problem 9.6). Alternatively, the order of the A and B partial steps can be interchanged at successive timesteps, thus mimicking Strang splitting.

Strang splitting is something of a chimera, since the source term B can often be incorporated directly into a high-order timestepping scheme for the operator A, and even when not, modern methods always use dynamic timesteps, so the effect of (9.8) is achieved due to the variable Δt's without having to take a first and last half timestep with A.

Timestep splitting for two or more spatial operators may also introduce second-order global errors even with Strang splitting or variable timesteps. For example, in solving the linear advection-diffusion equation with variable diffusion coefficient $D(x)$

$$u_t = -cu_x + (D(x)u_x)_x \equiv (A+B)u \tag{9.9}$$

we could update with a hyperbolic method for advection

$$u^* = e^{\Delta t A} u^n \tag{9.10}$$

and a parabolic method for diffusion

$$u^{n+1} = e^{\Delta t B} u^*. \tag{9.11}$$

Without the Strang technique, timestep splitting will again introduce locally second-order and globally first-order errors if the commutator of A with B

$$[A, B] = -c(D_x u_x)_x \neq 0 \tag{9.12}$$

independent of the spatial and temporal orders of accuracy of the methods for each partial timestep. With Strang splitting or dynamic timestepping, splitting still introduces second-order errors globally. Of course if $[A, B] = 0$ (and boundary conditions are handled appropriately), we recover the orders of accuracy of the A and B methods.

9.2 Compressible Navier-Stokes Equations

The most general classical fluid dynamical equations are the compressible Navier-Stokes equations, which consist of conservation laws for mass (9.13), momentum (9.14), and energy (9.15) for a compressible fluid with heat conduction, viscosity, and external forces:

$$\frac{\partial \rho}{\partial t} + \frac{\partial}{\partial x_i}(\rho u_i) = 0 \tag{9.13}$$

$$\frac{\partial}{\partial t}(\rho u_j) + \frac{\partial}{\partial x_i}(\rho u_i u_j - P_{ij}) = nF_j \tag{9.14}$$

$$\frac{\partial E}{\partial t} + \frac{\partial}{\partial x_i}(u_i E - u_j P_{ij} + q_i) = nu_i F_i \tag{9.15}$$

where ρ is the mass density, $\mathbf{u}$ is the velocity, P_{ij} is the stress tensor, $n = \rho/m$ is the number density with m the mass of the fluid particles, $\mathbf{F}$ represents the external forces acting on the fluid like gravity or an electric field, E is the energy density, and $\mathbf{q}$ is the heat flux. Indices i, j equal 1, 2, 3, and repeated indices are summed over.

The stress tensor is given by

$$P_{ij} = -P\delta_{ij} - \frac{2}{3}\mu \frac{\partial u_k}{\partial x_k}\delta_{ij} + \mu\left(\frac{\partial u_i}{\partial x_j} + \frac{\partial u_j}{\partial x_i}\right) \quad (9.16)$$

where P is the pressure (specified in terms of the other state variables ρ, $\mathbf{u}$, and E by an equation of state) and μ is the dynamic viscosity. Note that μ has units of $\rho\nu$, where the viscosity ν has units of l^2/t. The heat flux is usually approximated by the Fourier law $\mathbf{q} = -\kappa \nabla T$ ($\kappa > 0$), where T is the temperature.

The conservation laws can be derived from Newton's laws applied to a fluid element, or by a moment expansion of the Boltzmann transport equation (see [40], e.g.).

The incompressible Navier-Stokes equations (see Sect. 9.4) and the Euler equations of ideal gas dynamics (simply neglect heat conduction and viscosity) can be derived by making restrictions on these general equations. The drift-diffusion model below can also be derived (see Problem 9.2) from the compressible Navier-Stokes equations by adding a collision term $-\rho\mathbf{u}/\tau$ to the right-hand side of the momentum conservation equation (9.14), neglecting the energy conservation equation (9.15), and adding Poisson's equation for the electrostatic potential.

9.3 Drift-Diffusion Model

Consider a flow of positive and negative ions (cations and anions) in water for biological cells plus surrounding baths in an electric field $\mathbf{E}(\mathbf{x}, t)$. Model the discrete distribution of charges by continuum particle densities $n_\alpha(\mathbf{x}, t)$, $\alpha = 1, 2, \ldots, N_\alpha$ for the mobile ions (K^+, Cl^-, Na^+, Ca^{++}, ...) and $N_m(\mathbf{x})$ for fixed ions. Note that here N_m vanishes by definition in the baths and inside cells; N_m is only nonzero for the membrane surface charges σ_α in this approximation.

Mathematically, the flow of ions can be modeled by the drift-diffusion (or Poisson-Nernst-Planck) equations—partial differential equations for conservation of each ion species and Poisson's equation for the electrostatic potential $\phi(\mathbf{x}, t)$:

$$\frac{\partial n_\alpha}{\partial t} + \nabla \cdot (q_\alpha \mu_\alpha \mathbf{E} n_\alpha) = \nabla \cdot (D_\alpha \nabla n_\alpha), \quad \alpha = 1, 2, \ldots, N_\alpha \quad (9.17)$$

$$\nabla \cdot (\varepsilon \nabla \phi) = -\left(eN_m + \sum_\alpha q_\alpha n_\alpha\right), \quad \mathbf{E} = -\nabla \phi. \quad (9.18)$$

where q_α is the ionic charge, μ_α is the usual mobility coefficient divided by the electronic charge $e > 0$, D_α is the diffusion coefficient, and ε is the permittivity. A repeated index α is *not* summed over unless there is an explicit summation sign $\sum_\alpha$.

9.3 Drift-Diffusion Model

The ion current density is

$$\mathbf{j}_\alpha = q_\alpha \mu_\alpha n_\alpha \mathbf{E} - D_\alpha \nabla n_\alpha \quad (9.19)$$

for each ion species α, and the total electric current density is

$$\mathbf{j}_{\text{electric}} = \sum_\alpha q_\alpha \mathbf{j}_\alpha. \quad (9.20)$$

The physical parameters ε, μ_α, and D_α are functions of $\mathbf{x}$. The mobility coefficient μ_α and the diffusion coefficients D_α obey Einstein's relation $D_\alpha/\mu_\alpha = k_B T$.

The drift-diffusion equations form a parabolic/elliptic system of PDEs: The transport equation (9.17) is parabolic, and Poisson's equation (9.18) is elliptic. Thus, the boundary conditions for both n_α and ϕ are Dirichlet and/or Neumann.

In [14], we applied the drift-diffusion model to the triad synapse of the retina (see Fig. 9.1). Since biologists are primarily interested in steady-state current-voltage curves for the triad synapse, we simulate until steady state is reached by requiring

$$\left\| \frac{dn}{dt} \right\|_1 \equiv \frac{1}{N_\alpha (N_x - 1)(N_y - 1)} \sum_{\alpha=1}^{N_\alpha} \sum_{i=1}^{N_x-1} \sum_{j=1}^{N_y-1} \left| \frac{\Delta n_{\alpha ij}}{\Delta t} \right| \leq \epsilon \quad (9.21)$$

where N_x is the number of Δx, N_y is the number of Δy, and $\epsilon = 10^{-6}$, for example.

Fig. 9.1 Simulation by C. L. Gardner, J. R. Jones, S. M. Baer, and S. M. Crook [14] of steady-state electric potential in mV (from 0 mV at the top to 60 mV at the bottom left) in the triad synapse of the retina on a $600\Delta x \times 900\Delta y$ = 600 nm × 900 nm grid. The cone pedicle is at the top, the horizontal cell is the upward spike from the bottom, and the bipolar cell is at the left side

The steady-state boundary condition types for cell membranes plus extra- and intracellular baths with no transmembrane currents are defined by

$$n_\alpha = n_{b\alpha}, \quad \phi = 0 \quad \text{(bath far-field BC)} \tag{9.22}$$

$$n_\alpha = n_{b\alpha}, \quad \hat{\mathbf{n}} \cdot \nabla\phi = 0 \quad \text{(ambient bath BC)} \tag{9.23}$$

$$\hat{\mathbf{n}} \cdot \nabla n_\alpha = 0, \quad \hat{\mathbf{n}} \cdot \nabla\phi = 0 \quad \text{(membrane no-flux BC)} \tag{9.24}$$

where $\hat{\mathbf{n}}$ is a unit normal vector to the boundary. Membrane boundary conditions with transmembrane currents are discussed below.

Poisson's equation enforces charge neutrality away from membrane boundaries over any small but macroscopic length scale on the order of a few Debye lengths or larger, where the Debye length for ions in the intra- and extracellular baths is

$$l_D = \sqrt{\frac{\varepsilon k_B T}{\sum_\alpha q_\alpha^2 n_{b\alpha}}} \approx 1 \text{ nm}. \tag{9.25}$$

Variables $n_{\alpha ij}$ and ϕ_{ij} are defined at grid points $i = 0, 1, \ldots, N_x$ and $j = 0, 1, \ldots, N_y$, while $\mathbf{j}_{\text{electric}}$ and $\mathbf{E}$ are defined at the midpoints of grid elements connecting the grid points. Given n_α^n and ϕ^n at time level n, a timestep consists of three partial steps:

- First solve the transport equation (9.17) for n_α^{n+1}, $\alpha = 1, 2, \ldots, N_\alpha$, using TRBDF2 with $\phi = \phi^n$.
- Then solve Poisson's equation (9.18) for ϕ^{n+1} using PCG with n_α^{n+1} on the right-hand side.
- Finally update the membrane surface charges $\sigma_\alpha^\pm$, $\alpha = 1, 2, \ldots, N_\alpha$ through the membrane boundary conditions (9.27)–(9.29).

The membrane is modeled as a surface of zero thickness, with ionic surface charge densities $\sigma_\alpha^\pm$ and electrostatic potential $\phi^\pm$ on either side. The two sides of the membrane are labeled as "+" (intracellular) and "−" (extracellular). We assume that the membrane is charge neutral:

$$\sum_\alpha \sigma_\alpha^+ \equiv \sigma = -\sum_\alpha \sigma_\alpha^-. \tag{9.26}$$

The membrane boundary conditions are

$$\frac{\partial \sigma_\alpha^\pm}{\partial t} = \mp \hat{\mathbf{n}} \cdot (q_\alpha \mathbf{j}_\alpha^\pm) \pm j_{m\alpha}^\pm \tag{9.27}$$

$$V \equiv [\phi] = \phi^+ - \phi^- = \frac{\sigma}{C_m} \tag{9.28}$$

$$[\hat{\mathbf{n}} \cdot \nabla \phi] = 0 \qquad (9.29)$$

where $\hat{\mathbf{n}}$ is a unit normal vector to the boundary pointing from the "−" side to the "+" side of the membrane, V is the membrane potential, and C_m is the membrane capacitance per unit area. The transmembrane current densities $j_{m\alpha}(\mathbf{x}, t)$ (in the normal direction to the membrane) are defined by Hodgkin-Huxley phenomenological expressions.

When solving the drift-diffusion equations near the membrane boundary, we need to convert the surface charge densities σ_α into equivalent space charge densities $n_{m\alpha}$ over a boundary layer near the membrane:

$$q_\alpha n_{m\alpha} = q_\alpha n_{b\alpha} + \frac{\sigma_\alpha}{l_D}. \qquad (9.30)$$

Equation (9.30) can be derived near thermal equilibrium from the Poisson-Boltzmann equation

$$\nabla \cdot (\varepsilon \nabla \phi) = -\left(eN_m + \sum_\alpha q_\alpha n_{b\alpha} \exp\left\{-\frac{q_\alpha \phi}{k_B T}\right\}\right). \qquad (9.31)$$

9.4 Incompressible Navier-Stokes Equations

Incompressible flow invokes just the first two conservation laws for mass (9.13) and momentum (9.14) from the compressible Navier-Stokes equations and is characterized by setting the total time derivative of the density equal to zero:

$$\frac{d\rho}{dt} = \frac{\partial \rho}{\partial t} + \mathbf{u} \cdot \nabla \rho = 0 \qquad (9.32)$$

or

$$\nabla \cdot \mathbf{u} = 0 \qquad (9.33)$$

by virtue of (9.13). In other words, incompressibility means that the density in a fluid is constant along particle paths. An important special case is when the density $\rho = \rho_0$ is everywhere constant, which we will henceforth assume. The incompressible Navier-Stokes equations are then written as

$$\frac{\partial \mathbf{u}}{\partial t} + \mathbf{u} \cdot \nabla \mathbf{u} = -\nabla P + \nu \nabla^2 \mathbf{u} \qquad (9.34)$$

$$\nabla \cdot \mathbf{u} = 0 \qquad (9.35)$$

where P is the original pressure divided by ρ_0. The incompressible Navier-Stokes equations are called incompletely parabolic, since the parabolic transport equation (9.34) is coupled with the elliptic velocity constraint (9.35).

Chorin [6] introduced an elegant method for computationally solving the Navier-Stokes equations by combining a velocity update neglecting P (similar to the viscous Burgers equation) with a pressure correction and a divergence-free velocity projection (resulting in a Poisson equation).

We always use dimensionless units for computations, so defining the (dimensionless) Reynolds number $Re = Ud/\nu$ where U is a typical velocity scale and d is a typical length scale in the problem, we obtain (after multiplying (9.34) by d/U^2 and (9.35) by d/U) the Navier-Stokes equations in dimensionless form:

$$\frac{\partial \mathbf{u}}{\partial t} + \mathbf{u} \cdot \nabla \mathbf{u} = -\nabla P + \frac{1}{Re} \nabla^2 \mathbf{u} \tag{9.36}$$

$$\nabla \cdot \mathbf{u} = 0 \tag{9.37}$$

where $\mathbf{u}$ and P as well as ∂_t and ∇ are now dimensionless.

Chorin's method begins with a partial timestep updating the velocity while ignoring the pressure term, yielding a 3D viscous Burgers equation, using the appropriate Dirichlet or Neumann fluid dynamical boundary conditions for the velocity:

$$\frac{\mathbf{u}^* - \mathbf{u}^n}{\Delta t} = -\mathbf{u}^n \cdot \nabla \mathbf{u}^n + \frac{1}{Re} \nabla^2 \mathbf{u}^n. \tag{9.38}$$

Then the pressure term is included via a partial timestep

$$\frac{\mathbf{u}^{n+1} - \mathbf{u}^*}{\Delta t} = -\nabla P^{n+1} \tag{9.39}$$

again using the fluid dynamical boundary conditions for the velocity, with the incompressibility condition

$$\nabla \cdot \mathbf{u}^{n+1} = 0 \tag{9.40}$$

imposed on the new velocity. In (9.38) and (9.39), $\Delta \mathbf{u}/\Delta t$ is a shorthand for any consistent (explicit) timestepping scheme like second-order predictor-corrector or RK4.

To solve for the new pressure P^{n+1} in (9.39), we take the divergence of (9.39) to obtain a Poisson equation

$$\nabla^2 P^{n+1} = \frac{1}{\Delta t} \nabla \cdot \mathbf{u}^* \tag{9.41}$$

with numerical Neumann boundary condition from (9.39)

9.4 Incompressible Navier-Stokes Equations

$$\hat{\mathbf{n}} \cdot \nabla P_{\mathcal{B}}^{n+1} = -\frac{1}{\Delta t} \hat{\mathbf{n}} \cdot \left(\mathbf{u}_{\mathcal{B}}^{n+1} - \mathbf{u}_{\mathcal{B}}^{*} \right) \tag{9.42}$$

where $\hat{\mathbf{n}}$ is a unit outward-pointing normal vector to the boundary $\mathcal{B}$. As shown directly below, the right-hand side can be taken to be zero:

$$\hat{\mathbf{n}} \cdot \nabla P_{\mathcal{B}}^{n+1} = 0 \tag{9.43}$$

giving a numerical homogeneous Neumann boundary condition on the pressure. The Poisson equation (9.41) can be solved with PCG or a sparse direct method.

To validate (9.43), we can without loss of generality focus on the left boundary of the 2D problem with grid points $i = 0, 1 \ldots, N_x$ and $j = 0, 1, \ldots, N_y$. Pressures are defined at grid points, and velocities are defined at the midpoints of elements connecting grid points. Place the left boundary at grid point $\mathcal{B} = \left(\frac{1}{2}, m \right)$, where m is an integer between 1 and $N_y - 1$. Then discretizing (9.41), we obtain

$$\frac{1}{\Delta x^2} \left(P_{2,m}^{n+1} - 2 P_{1,m}^{n+1} + P_{0,m}^{n+1} \right) + \frac{1}{\Delta y^2} \left(P_{1,m+1}^{n+1} - 2 P_{1,m}^{n+1} + P_{1,m-1}^{n+1} \right) =$$

$$\frac{1}{\Delta t} \left(\frac{1}{\Delta x} \left(u_{\frac{3}{2},m}^{*} - u_{\mathcal{B}}^{*} \right) + \frac{1}{\Delta y} \left(v_{1,m+\frac{1}{2}}^{*} - v_{1,m-\frac{1}{2}}^{*} \right) \right) \tag{9.44}$$

where $\mathbf{u} = (u, v)$ in 2D. Next discretize (9.42) at the boundary point $\left(\frac{1}{2}, m \right)$:

$$\frac{1}{\Delta x} \left(P_{1,m}^{n+1} - P_{0,m}^{n+1} \right) = -\frac{1}{\Delta t} \left(u_{\mathcal{B}}^{n+1} - u_{\mathcal{B}}^{*} \right). \tag{9.45}$$

Now when (9.45) is substituted into (9.44), $u_{\mathcal{B}}^{*}$ disappears from the problem, so we can choose the numerical boundary condition $u_{\mathcal{B}}^{*} = u_{\mathcal{B}}^{n+1}$ and the right-hand side of (9.45) is zero.

Thus Chorin's method involves the following three partial steps:

- Solve the 3D viscous Burgers equation (9.38) for $\mathbf{u}^{*}$ using (e.g.) second-order predictor-corrector, with the appropriate Dirichlet or Neumann fluid dynamical boundary conditions for the velocity:

$$\frac{\mathbf{u}^{*} - \mathbf{u}^{n}}{\Delta t} = -\mathbf{u}^{n} \cdot \nabla \mathbf{u}^{n} + \frac{1}{Re} \nabla^2 \mathbf{u}^{n}.$$

- Solve the Poisson equation (9.41) for P^{n+1} using (e.g.) PCG, with the homogeneous Neumann boundary condition (9.43):

$$\nabla^2 P^{n+1} = \frac{1}{\Delta t} \nabla \cdot \mathbf{u}^{*}, \quad \hat{\mathbf{n}} \cdot \nabla P_{\mathcal{B}}^{n+1} = 0.$$

- Solve (9.39) for $\mathbf{u}^{n+1}$ using second-order predictor-corrector, with the fluid dynamical boundary conditions for the velocity:

$$\frac{\mathbf{u}^{n+1} - \mathbf{u}^*}{\Delta t} = -\nabla P^{n+1}.$$

For a project on the 2D Navier-Stokes equations, see Problem 9.7.

9.5 Classical Hydrodynamic Model

The classical hydrodynamic (CHD) model (electro-gas dynamics with heat conduction) for semiconductor devices consists of the following conservation laws:

$$\frac{\partial n}{\partial t} + \frac{\partial}{\partial x_i}(nu_i) = 0 \qquad (9.46)$$

$$\frac{\partial}{\partial t}(mnu_j) + \frac{\partial}{\partial x_i}(mnu_i u_j) = -\frac{\partial P}{\partial x_j} - n\frac{\partial V}{\partial x_j} - \frac{mnu_j}{\tau_p} \qquad (9.47)$$

$$\frac{\partial W}{\partial t} + \frac{\partial}{\partial x_i}(u_i(W+P) + q_i) = -nu_i\frac{\partial V}{\partial x_i} - \frac{\left(W - \frac{3}{2}nk_B T_0\right)}{\tau_w} \qquad (9.48)$$

$$\nabla \cdot (\varepsilon \nabla V) = e^2(N_d - n) \qquad (9.49)$$

where n is the electron number density (we could also add a set of equations for holes), $\mathbf{u}$ is the velocity, m is the effective mass, $P = nk_B T$ is the pressure, T is the temperature, $V = -e\phi$ is the electrostatic potential energy, $e > 0$ is the electronic charge, ϕ is the electrostatic potential with electric field $\mathbf{E} = -\nabla\phi = \nabla(V/e)$, $W = \frac{3}{2}nk_B T + \frac{1}{2}mnu^2$ is the energy density[1] ($\gamma = 5/3$), $\mathbf{q} = -\bar{\kappa}n\nabla T$ is the heat flux ($\bar{\kappa}$ is a constant), and T_0 is the temperature of the semiconductor lattice. Indices i, j equal 1, 2, 3, and repeated indices are summed over. Equation (9.46) expresses conservation of electron number, Eq. (9.47) conservation of momentum, and Eq. (9.48) conservation of energy. The last terms in (9.47) and (9.48) represent electron scattering, which is modeled by the standard relaxation time approximation, with momentum and energy relaxation times $\tau_p = \tau_p(T)$ and $\tau_w = \tau_w(T)$. In Poisson's equation (9.49), $N_d = N_d(\mathbf{x})$ is the background doping density in the semiconductor device.

---

[1] We use W instead of E for the energy density in the CHD model, to avoid confusion with the electric field.

9.5 Classical Hydrodynamic Model

The 1D CHD model PDEs have two hyperbolic, two parabolic, and two elliptic modes (see Problems 9.3 and 9.4). The CHD code **chd.m** uses WENO3 for the hyperbolic transport modes, TRBDF2 for the parabolic heat conduction mode, and a tridiagonal direct solve for the elliptic Poisson equation.

A timestep from time level n to time level $n + 1$ consists of the following partial steps:

- Solve the gas dynamical hyperbolic transport equations (9.46)–(9.48) with $\bar{\kappa} = 0$ using WENO3 to obtain n^{n+1}, $(\rho u)^{n+1}$, and $\widetilde{W}^{n+1}$, with τ_p, τ_w, and V source terms (all at time level n) included in the RK3 method.
- Update the temperature $\widetilde{T}^{n+1}$ via

$$\frac{\partial T}{\partial t} = \frac{2}{3nk_B} \frac{\partial}{\partial x}\left(\bar{\kappa} n \frac{\partial T}{\partial x}\right)$$

from (9.48) using TRBDF2 to obtain T^{n+1} and then correct W^{n+1} with the new T^{n+1}.
- Use a tridiagonal direct solve for Poisson's equation (9.49) to obtain the electrostatic potential energy V^{n+1}.

Since semiconductor device engineers are primarily interested in steady-state current-voltage curves, we typically simulate until steady state is reached by requiring

$$\left\|\frac{dq}{dt}\right\|_1 \equiv \frac{1}{3(N-1)} \sum_{\alpha=1}^{3} \sum_{i=1}^{N-1} \left|\frac{\Delta q_{\alpha i}}{\Delta t}\right| \leq \epsilon \tag{9.50}$$

where $\epsilon = 10^{-6}$, for example, and the conserved variables are

$$q_\alpha = \begin{bmatrix} n \\ \rho u \\ W \end{bmatrix}. \tag{9.51}$$

Figures 9.2, 9.3, 9.4 and 9.5 illustrate the development of a steady-state electron shock wave [13] in an n^+-n-n^+ diode. The diode consists of an n^+ (electron rich) source, an n (still electron rich, but less so) channel, and an n^+ (electron rich) drain, and models electron flow in the channel of a field effect transistor.

Six boundary conditions (two for the hyperbolic transport modes with subsonic outflow), two for parabolic heat conduction, and two for the elliptic Poisson equation) are imposed at the left/right boundaries of the device: $V_{\text{left}} = 0$, $V_{\text{right}} = -V_{\text{bias}}$ (so that electrons flow from left to right for positive voltage biases), $n = N$ left and right, $dT/dx = 0$ left and right.

In 1D steady state, (9.46) implies $nu = \text{const}$. The background doping N_d in the diode plays the role of the converging/diverging geometry in a de Laval nozzle: Electron density decreases in the channel (Fig. 9.2), causing the velocity

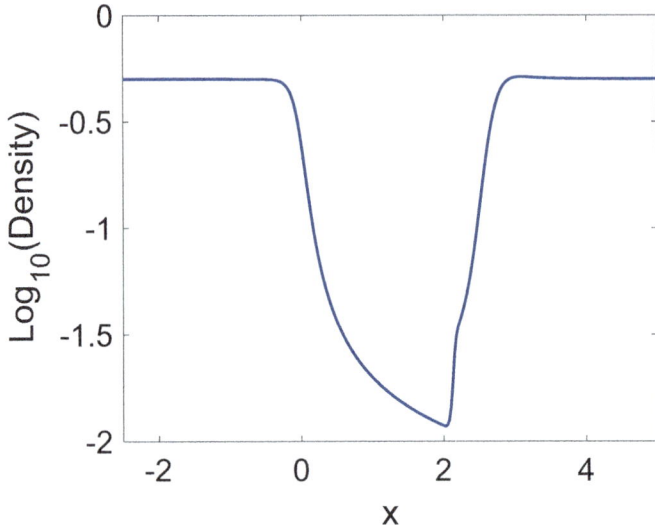

Fig. 9.2 WENO3 simulation to steady state of $\log_{10}(n/\bar{n})$, where $\bar{n} = 10^{18}$ electrons/cm^3, depicting a Mach 3.2 electron shock wave in a 0.25 micron channel GaAs diode using **chd.m** with $300\Delta x$ and $r = 0.9$

u to increase dramatically to supersonic flow (Fig. 9.3). Then as electrons enter the drain, density increases (to preserve near charge neutrality), and the electrons slow down to subsonic velocities (and heat up to $T \approx 10\,T_{\text{amb}}$ in Fig. 9.4). A shock wave is generated as the electrons transition from supersonic to subsonic flow. Figure 9.5 shows small potential barriers (within the overall decrease in V) at the junctions due to dipole ($+/-$ charge distribution) layers.

A shock wave with heat conduction and no viscosity has a more complicated structure [10, 42] than the exact discontinuity in ideal gas dynamics: There is an outer smooth profile to a shock wave, with (above a critical Mach number $M_c \approx 1.7$ for an electron gas with $\gamma = 5/3$) an inner exact discontinuity (see Fig. 9.3).

The development of the electron shock wave in a one micron Si semiconductor device at 77 K was verified by a Monte Carlo simulation of the Boltzmann transport equation in [13].

Problems

9.1 Derive the incompressible Navier-Stokes equations from the compressible Navier-Stokes equations (see Sects. 9.2 and 9.4) by (i) setting $d\rho/dt = \rho_t + \mathbf{u} \cdot \nabla \rho = 0$ (incompressibility means ρ is constant along particle paths) in the conservation of mass equation (9.13), (ii) neglecting the conservation of energy equation (9.15), and (iii) defining $\mu = \rho \nu$ and simplifying the conservation of momentum equation

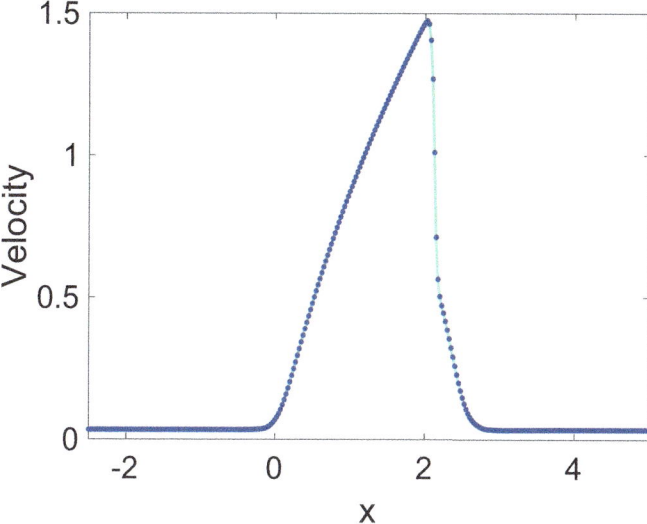

Fig. 9.3 WENO3 simulation to steady state of velocity in 10^8 cm/s depicting a Mach 3.2 electron shock wave using **chd.m**. The *dots* are grid point values. With heat conduction but no viscosity, the shock has an inner discontinuity and an outer smooth profile

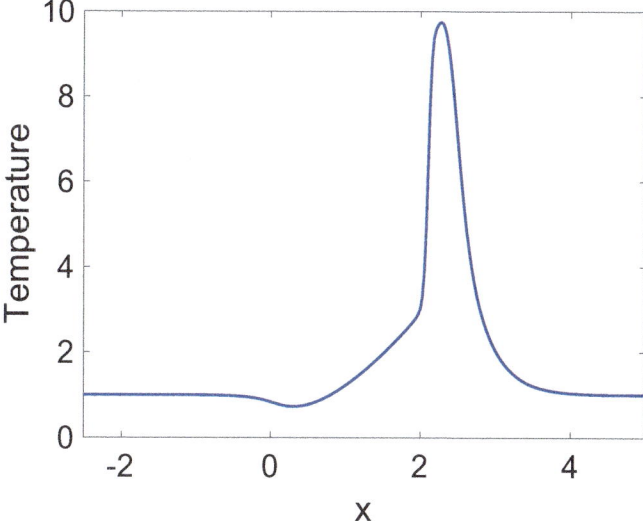

Fig. 9.4 WENO3 simulation to steady state of temperature T/T_{amb}, where $T_{amb} = 300$ K is the ambient temperature, using **chd.m**

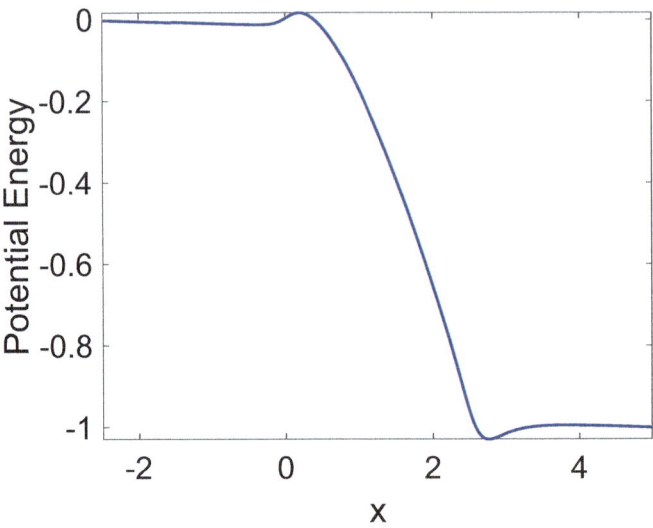

Fig. 9.5 WENO3 simulation to steady state of the electrostatic potential energy in eV using **chd.m**

(9.14) with

$$P_{ij} = -P\delta_{ij} - \frac{2}{3}\mu \frac{\partial u_k}{\partial x_k}\delta_{ij} + \mu\left(\frac{\partial u_i}{\partial x_j} + \frac{\partial u_j}{\partial x_i}\right).$$

Note that since $\nabla \cdot \mathbf{u} = 0$,

$$P_{ij} = -P\delta_{ij} + \mu\left(\frac{\partial u_i}{\partial x_j} + \frac{\partial u_j}{\partial x_i}\right).$$

9.2 Derive the drift-diffusion model equations in Sect. 9.3 from the compressible Navier-Stokes equations in Sect. 9.2 (it is sufficient to consider just a single ion species): Neglect the energy conservation equation (9.15), so that T is constant; set $\mathbf{F} = q\mathbf{E}$ in the momentum conservation equation (9.14); and add Poisson's equation for the electrostatic potential. Then

(a) Derive the particle flux $\mathbf{j} = n\mathbf{u}$ by adding a collision term $-\rho\mathbf{u}/\tau$ to the right-hand side of the momentum conservation equation, where τ is a relaxation time, and setting $\partial_t = 0$. *Hint:* Choose τ appropriately.
(b) Then plug $\mathbf{j}$ into the conservation of particles equation (this is the conservation of mass equation (9.13) divided by m)

$$\frac{\partial n}{\partial t} + \nabla \cdot \mathbf{j} = 0$$

to derive (9.17).

9.3 For the 1D classical hydrodynamic model, show the eigenvalues σ of the symbol S of the linearized PDE system (decouple the elliptic Poisson equation) are

$$\sigma_0 \approx \frac{2\bar{\kappa}}{3}k^2, \quad \text{2 hyperbolic}$$

$$\sigma_\pm = ik(u \pm c), \quad \text{1 parabolic}$$

where now $c = \sqrt{T/m}$. The eigenvalue iku and the associated contact wave have disappeared. *Hints:* Set $1/\tau_p = 0 = 1/\tau_w$. Use the physical variables $w = [n, u, T]$. Show that in terms of physical variables, the CHD equations become

$$\frac{\partial n}{\partial t} + \frac{\partial}{\partial x}(nu) = 0$$

$$\frac{\partial u}{\partial t} + u\frac{\partial u}{\partial x} + \frac{1}{mn}\frac{\partial}{\partial x}(nT) + \frac{1}{m}\frac{\partial V}{\partial x} = 0$$

$$\frac{\partial T}{\partial t} + u\frac{\partial T}{\partial x} + \frac{2T}{3}\frac{\partial u}{\partial x} - \frac{2}{3n}\frac{\partial}{\partial x}\left(\bar{\kappa}n\frac{\partial T}{\partial x}\right) = 0$$

$$-\frac{\partial^2 V}{\partial x^2} + \frac{e^2}{\epsilon}(N - n) = 0.$$

There is an elliptic k^2 mode from Poisson's equation plus the three modes of the three transport equations. Now derive the symbol S of the CHD transport equations:

$$S = \begin{bmatrix} iku & ikn & 0 \\ ik\frac{T}{mn} & iku & \frac{ik}{m} \\ 0 & \frac{2}{3}ikT & iku + \frac{2\bar{\kappa}}{3}k^2 \end{bmatrix}.$$

Then you can simply verify the Tr$\{S\}$ and det$\{S\}$ formulas for the eigenvalues.

9.4 The analysis in Problem 9.3 can be extended to the 3D CHD model (and 3D gas dynamics).

(a) Show the eigenvalues σ of the symbol S of the linearized 3D CHD system (decouple the elliptic Poisson equation) are

$$i(\mathbf{k} \cdot \mathbf{u} \pm kc), \quad \text{2 hyperbolic}$$

$$i\mathbf{k} \cdot \mathbf{u}, \quad \text{2 hyperbolic}$$

$$\frac{2\bar{\kappa}}{3}k^2, \quad \text{1 parabolic}$$

where $c = \sqrt{T/m}$, for a total of four hyperbolic modes and one parabolic mode plus the elliptic Poisson equation mode. There are two families of shock waves with characteristic velocities $u_n \pm c$, where u_n is the velocity normal to the wave, and two families of contact waves with characteristic velocity u_n corresponding to jumps in tangential velocity. The contact wave corresponding to a discontinuity in T has disappeared due to the parabolic heat conduction term.

(b) Now set $\bar{\kappa} = 0$ (electro-gas dynamics without heat conduction). Show the eigenvalues σ of the symbol S of the linearized system (decouple the elliptic Poisson equation) are

$$i(\mathbf{k} \cdot \mathbf{u} \pm kc), \quad \text{2 hyperbolic}$$

$$i\mathbf{k} \cdot \mathbf{u}, \quad \text{3 hyperbolic}$$

where $c = \sqrt{5T/(3m)}$, for a total of five hyperbolic modes plus the elliptic Poisson equation mode. There are two families of shock waves with characteristic velocities $u_n \pm c$ and three families of contact waves with characteristic velocity u_n corresponding to jumps in tangential velocity and T.

Hints: Set $1/\tau_p = 0 = 1/\tau_w$. Use the physical variables $w = [n, \mathbf{u}, T]$. Then you can simply verify the Tr$\{S\}$ and det$\{S\}$ formulas for the eigenvalues.

9.5 Show **chd.m** converges under mesh refinement by simulating the electron shock wave in a GaAs 0.25 micron channel diode at 300 K with $N = 150, 300$, and $600\Delta x$. Use

chd(5,1,150,1)

etc. Compare the steady-state velocity plots for each of three cases. Discuss your results.

9.6 In the error for Strang splitting, calculate the $O(\Delta t^3)$ term.

9.7 *Project:* Develop a 2D Navier-Stokes solver for lid-driven cavity flow (see Sect. 6.7 of Strang's *Computational Science and Engineering* [36]). The physical problem is a unit square filled with fluid with velocity boundary conditions $(u, v) = (0, 0)$ on three sides and $(u, v) = (1, 0)$ on the top. The fluid rotates within the square, with counterrotating vortices. Explore the pattern of vortices vs. Reynolds number $Re = Ud/\nu = 1/\nu$, where U is a typical velocity scale and d is a typical length scale in the problem.

Appendix A
Useful Mathematical Formulas

A.1 MATLAB vs. C Indices

In the text, we will index quantities based on clarity and simplicity, usually following C conventions. For MATLAB instead of C, simply shift the index $i \to i + 1$.

For example, in 1D space with $x \in [a, b]$, we discretize in C (or C++, etc.) on a spatial grid with fixed steps $\Delta x = (b - a)/N$ and grid points $x_0 = a, x_1, \ldots, x_N = b$. In MATLAB, we label grid points $x_1 = a, x_2, \ldots, x_N, x_{N+1} = b$.

As another example, solution values are labeled $u_0, u_1, \ldots, u_N$ in C and $u_1, u_2, \ldots, u_{N+1}$ in MATLAB.

A.2 Taylor Series Formulas

A.2.1 Taylor Series for $f(x)$

$$f(x+\Delta x) = f(x) + \Delta x\, f'(x) + \frac{\Delta x^2}{2!} f''(x) + \frac{\Delta x^3}{3!} f'''(x) + \cdots = \sum_{n=0}^{\infty} \frac{\Delta x^n}{n!} f^{(n)}(x)$$

$$f(x - \Delta x) = f(x) - \Delta x\, f'(x) + \frac{\Delta x^2}{2!} f''(x) - \frac{\Delta x^3}{3!} f'''(x) + \cdots$$
$$= \sum_{n=0}^{\infty} (-1)^n \frac{\Delta x^n}{n!} f^{(n)}(x)$$

A.2.2 Taylor Series for $u(t)$

$$u(t + \Delta t) = u(t) + \Delta t\, u'(t) + \frac{\Delta t^2}{2!} u''(t) + \frac{\Delta t^3}{3!} u'''(t) + \cdots = \sum_{n=0}^{\infty} \frac{\Delta t^n}{n!} u^{(n)}(t)$$

A.2.3 Taylor Series for $f(x, y)$

$$f(x + \Delta x, y + \Delta y) = f(x, y) + \Delta x\, f_x(x, y) + \Delta y\, f_y(x, y)$$

$$+ \frac{\Delta x^2}{2!} f_{xx}(x, y) + \Delta x \Delta y\, f_{xy}(x, y) + \frac{\Delta y^2}{2!} f_{yy}(x, y) + \cdots$$

$$= \sum_{m,n=0}^{\infty} \frac{\Delta x^m \Delta y^n}{m!n!} \frac{\partial^{m+n} f}{\partial x^m \partial y^n}(x, y)$$

A.2.4 Taylor Series for $u(x, t)$

$$u(x + \Delta x, t + \Delta t) = u(x, t) + \Delta x\, u_x(x, t) + \Delta t\, u_t(x, t)$$

$$+ \frac{\Delta x^2}{2!} u_{xx}(x, t) + \Delta x \Delta t\, u_{xt}(x, t) + \frac{\Delta t^2}{2!} u_{tt}(x, t) + \cdots$$

$$= \sum_{m,n=0}^{\infty} \frac{\Delta x^m \Delta t^n}{m!n!} \frac{\partial^{m+n} u}{\partial x^m \partial t^n}(x, t)$$

A.3 Basic Finite Difference Derivatives

A.3.1 Spatial Derivatives

Second-order accurate central difference approximation to first derivative:

$$\left(\frac{df}{dx}\right)_i \approx \frac{f_{i+1} - f_{i-1}}{2\Delta x} = f_i' + \frac{\Delta x^2}{6} f_i''' + \cdots$$

First-order accurate backward difference approximation to first derivative:

$$\left(\frac{df}{dx}\right)_i \approx \frac{f_i - f_{i-1}}{\Delta x} = f_i' - \frac{\Delta x}{2} f_i'' + \cdots$$

A Useful Mathematical Formulas

First-order accurate forward difference approximation to first derivative:

$$\left(\frac{df}{dx}\right)_i \approx \frac{f_{i+1} - f_i}{\Delta x} = f'_i + \frac{\Delta x}{2} f''_i + \cdots$$

Second-order accurate central difference approximation to second derivative:

$$\left(\frac{d^2 f}{dx^2}\right)_i \approx \frac{f_{i+1} - 2f_i + f_{i-1}}{\Delta x^2} = f''_i + \frac{\Delta x^2}{12} f_i^{(4)} + \cdots$$

A.3.2 Forward Time Difference

$$\frac{du}{dt} \approx \frac{u_{n+1} - u_n}{\Delta t} = \frac{u(t_{n+1}) - u(t_n)}{\Delta t} = \frac{u(t_n + \Delta t) - u(t_n)}{\Delta t} = u' + \frac{u''}{2} \Delta t + \cdots$$

With this formula, the forward and backward Euler methods for time-dependent differential equations are first-order accurate, while the TR method is second-order.

A.4 Orthogonality Relations for Fourier Series

For the Fourier sine and cosine series solutions, we make use of the orthogonality relations (m and n are positive (nonnegative) integers for sin (cos))

$$\int_0^\pi \sin(mx) \sin(nx) \, dx = \frac{\pi}{2} \delta_{mn}$$

$$\int_0^\pi \sin(mx) \cos(nx) \, dx = 0$$

$$\int_0^\pi \cos(mx) \cos(nx) \, dx = \frac{\pi c_n}{2} \delta_{mn}$$

where $c_0 = 2$ and $c_n = 1$, $n \geq 1$.

Appendix B
Norms and Condition Number

B.1 l_p Vector Norms for a Vector of Fixed Length

$$||x||_p = \left(\sum_{i=1}^{n} |x_i|^p\right)^{\frac{1}{p}}$$

$$||x||_1 = \sum_{i=1}^{n} |x_i|$$

$$||x||_2 = \left(\sum_{i=1}^{n} |x_i|^2\right)^{\frac{1}{2}} \quad \text{(Euclidean norm)}$$

$$||x||_\infty = \max_i |x_i|$$

B.2 l_p Vector Norms for a Vector of Variable Length n

For grid-based vectors of variable length n as we refine or coarsen the grid, we divide the $\sum_{i=1}^{n}$ by n:

$$||x||_p = \left(\frac{1}{n}\sum_{i=1}^{n} |x_i|^p\right)^{\frac{1}{p}}.$$

In particular, the 1-norm that we typically use in computations is

$$||x||_1 = \frac{1}{n}\sum_{i=1}^{n}|x_i|.$$

B.3 Basic Properties of Vector Norms

$$||x|| > 0 \quad \text{iff} \quad x \neq 0$$

$$||0|| = 0$$

$$||cx|| = |c|\,||x||, \quad c \text{ a scalar}$$

$$||x + y|| \leq ||x|| + ||y|| \quad \text{triangle inequality}$$

B.4 Matrix Norms

$$||A|| = \max \frac{||Ax||}{||x||}$$

The norm of a matrix depends on which vector norm is used. Note that $||My|| \leq ||M||\,||y||$.

$$||A^{-1}|| = \frac{1}{\min\{||Ax||/||x||\}}$$

There are simple formulas for the l_1, l_2, and l_∞ norms of a matrix:

$$||A||_1 = \max_j \sum_i |A_{ij}| \quad \text{(max column sum)}$$

$$||A||_\infty = \max_i \sum_j |A_{ij}| \quad \text{(max row sum)}$$

$$||A||_2 = \sqrt{\rho(A^\dagger A)}$$

where $\rho(B)$ denotes the spectral radius of the matrix B and $A^\dagger = (A^*)^T$. If A is a normal matrix, then $||A||_2 = \rho(A) = \max\{|\lambda|\}$.

B.5 Condition Number

$$\kappa(A) = ||A||\,||A^{-1}||$$

The condition number of a matrix also depends on which vector norm is used, but the various condition numbers are the same order of magnitude. For real symmetric matrices (and normal matrices in general),

$$||A||_2 = |\max\{|\lambda|\}|, \quad ||A^{-1}||_2 = \frac{1}{\min\{|\lambda|\}}, \quad \kappa(A) = \frac{\max\{|\lambda|\}}{\min\{|\lambda|\}}.$$

References

1. Bank, R.E., Coughran, W.M., Fichtner, W., Grosse, E.H., Rose, D.J., Smith, R.K.: Transient simulation of silicon devices and circuits. IEEE Transactions on Computer-Aided Design **4**, 436–451 (1985)
2. Bender, C.M., Orszag, S.A.: Advanced Mathematical Methods for Scientists and Engineers. Springer (1999)
3. Berger, M.J., Oliger, J.: Adaptive mesh refinement for hyperbolic partial differential equations. Journal of Computational Physics **53**, 484–512 (1984)
4. Briggs, W.L., Henson, V.E., McCormick, S.F.: A Multigrid Tutorial. SIAM (2000)
5. Butcher, J.C.: Numerical Methods for Ordinary Differential Equations. Wiley (2016)
6. Chorin, A.J.: Numerical solution of the Navier-Stokes equations. Mathematics of Computation **22**, 745–762 (1968)
7. Chorin, A.J.: Random choice solution of hyperbolic systems. Journal of Computational Physics **22**, 517–533 (1976)
8. Chorin, A.J., Marsden, J.E.: A Mathematical Introduction to Fluid Mechanics. Springer (1990)
9. Colella, P., Woodward, P.R.: The piecewise parabolic method (PPM) for gas-dynamical simulations. Journal of Computational Physics **54**, 174–201 (1984)
10. Courant, R., Friedrichs, K.O.: Supersonic Flow and Shock Waves. Springer (1976). Reprinted, originally published in 1948
11. Courant, R., Isaacson, E., Rees, M.: On the solution of nonlinear hyperbolic differential equations by finite differences. Communications on Pure and Applied Mathematics **5**, 243–255 (1952)
12. Du, J., Fix, B., Glimm, J., Jia, X., Li, X., Li, Y., Wu, L.: A simple package for front tracking. Journal of Computational Physics **213**, 613–628 (2006)
13. Gardner, C.L.: Hydrodynamic and Monte Carlo simulation of an electron shock wave in a one micrometer n+/n/n+ diode. IEEE Transactions on Electron Devices **40**, 455–457 (1993)
14. Gardner, C.L., Jones, J.R., Baer, S.M., Crook, S.M.: Drift-diffusion simulation of the ephaptic effect in the triad synapse of the retina. Journal of Computational Neuroscience **38**, 129–142 (2015)
15. Gardner, C.L., Jones, J.R., Scannapieco, E., Windhorst, R.A.: Numerical simulation of star formation by the bow shock of the Centaurus A jet. The Astrophysical Journal **835**, 232: 1–9 (2017)
16. George, A., Liu, J.W.H.: Computer Solution of Large Sparse Positive-Definite Systems. Prentice-Hall (1981)
17. Godunov, S.K.: A difference scheme for numerical solution of discontinuous solution of hydrodynamic equations. Matematicheskii Sbornik **47**, 271–306 (1959)

18. Gottlieb, D., Orszag, S.A.: Numerical Analysis of Spectral Methods. SIAM (1977)
19. Ha, Y., Gardner, C.L., Gelb, A., Shu, C.W.: Numerical simulation of high Mach number astrophysical jets with radiative cooling. Journal of Scientific Computing **24**, 29–44 (2005)
20. Hestenes, M.R., Stiefel, E.: Methods of conjugate gradients for solving linear systems. Journal of Research of the National Bureau of Standards **49**, 409–436 (1952)
21. Hu, X.Y., Adams, N.A., Shu, C.W.: Positivity-preserving method for high-order conservative schemes solving compressible Euler equations. Journal of Computational Physics **242**, 169–180 (2013)
22. Iserles, A.: A First Course in the Numerical Analysis of Differential Equations. Cambridge University Press (2009)
23. Johnson, M.J., Gardner, C.L.: An interface method for semiconductor process simulation. In: Semiconductors, *IMA Volumes in Mathematics and its Applications*, vol. 58, pp. 33–47. Springer (1993)
24. LeVeque, R.J.: Nonlinear conservation laws and finite volume methods for astrophysical fluid flow. In: Computational Methods for Astrophysical Fluid Flow, *Saas-Fee Advanced Course*, vol. 27, pp. 1–159. Springer (1998)
25. LeVeque, R.J.: Finite Volume Methods for Hyperbolic Problems. Cambridge University Press (2002)
26. LeVeque, R.J.: Finite Difference Methods for Ordinary and Partial Differential Equations. SIAM (2007)
27. Liska, R., Wendroff, B.: Comparison of several difference schemes on 1D and 2D test problems for the Euler equations. SIAM Journal on Scientific Computing **25**, 995–1017 (2003)
28. Lorenz, E.N.: Deterministic nonperiodic flow. Journal of Atmospheric Sciences **20**, 130–141 (1963)
29. Moler, C.: Numerical Computing with MATLAB. SIAM (2004)
30. Roe, P.L.: Approximate Riemann solvers, parameter vectors, and difference schemes. Journal of Computational Physics **135**, 250–258 (1997). Reprinted, originally published in 1981
31. Saad, Y., Schultz, M.H.: GMRES: A generalized minimal residual algorithm for solving nonsymmetric linear systems. SIAM Journal on Scientific and Statistical Computing **7**, 856–869 (1986)
32. Sethian, J.A.: Level Set Methods and Fast Marching Methods. Cambridge University Press (1999)
33. Shu, C.W.: High order ENO and WENO schemes for computational fluid dynamics. In: High-Order Methods for Computational Physics, *Lecture Notes in Computational Science and Engineering*, vol. 9, pp. 439–582. Springer (1999)
34. Smoller, J.: Shock Waves and Reaction-Diffusion Equations. Springer (1994)
35. Strang, G.: Introduction to Applied Mathematics. Wellesley-Cambridge Press (1986)
36. Strang, G.: Computational Science and Engineering. Wellesley-Cambridge Press (2009)
37. Taflove, A., Hagness, S.C.: Computational Electrodynamics: The Finite-Difference Time-Domain Method. Artech House (2005)
38. Toro, E.F.: Riemann Solvers and Numerical Methods for Fluid Dynamics. Springer (2009)
39. Trefethen, L.N., Bau, D.: Numerical Linear Algebra. SIAM (1997)
40. Whitham, G.B.: Linear and Nonlinear Waves. Wiley-Interscience (1974)
41. Young, D.M.: Iterative Solution of Large Linear Systems. Academic Press (1971). Reprinted by Dover, 2003
42. Zeldovich, Y.B., Raizer, Y.P.: Physics of Shock Waves and High-Temperature Hydrodynamic Phenomena. Dover (2002)

Index

A
Absolute stability, 17
Acronyms, xvii
Adaptive mesh refinement, 49
Advection equation, 121, 124
Arnoldi iteration algorithm, 112
A-stability, 16, 17, 22, 40, 42, 62, 67, 76
    trapezoidal rule, 24
Autonomous form, 26

B
Backward difference
    first derivative, 11, 202
Backward Euler method
    Newton's method, 39
    ODE IVPs, 22
    parabolic PDEs, 73
Backward heat/diffusion equation, 69, 70
Banded matrix direct solve, *see* Sparse matrix methods
Boundary conditions
    Cauchy, 58
    Dirichlet, 58, 77
    Neumann, 58, 66, 77
        homogeneous, 66
    parabolic/elliptic PDEs, 77
    periodic, 157
    Robin, 58
    through-flow, 157
    wall, 158
Boundary value problem, 49
    linear, 50
    nonlinear, 52
    numerical methods, 49
    order of accuracy, 51
Box method, *see* Finite volume methods
Burgers' equation, 121, 140
    boundary conditions, 145
    conservation form, 142
    inviscid, 141
    nonconservative shock, 146
    numerical methods, 145
    N-wave, 144, 148
    rarefaction, 140, 143, 148
    Riemann problem, 140
    shock, 140, 143, 146
    viscous, 140
    viscous shock profile, 143
    WENO3 code, 146

C
Cayley-Hamilton Theorem, 109
Central difference
    first derivative, 11, 202
    second derivative, 11, 203
CFL condition, 28, 68, 123, 130
CFL factor, 124
Chaotic relaxation, 104
Characteristic equations, 127, 153
Characteristics, 58, 121, 125, 126, 128, 139, 140, 142, 144, 154, 155, 157, 158, 172, 200
Characteristic variables, 127, 128, 163
Chorin's method, 192, 193
Classical hydrodynamic (CHD) model, 194
    classification, 199

Classical hydrodynamic (CHD) model (*cont.*)
  conserved variables, 195
  contact, 200
  electron shock, 195, 196, 200
  numerical methods, 195
  WENO3 code, 194
Classification
  linear PDEs, 58
  nonlinear systems of PDEs, 60
Code development, 6
Codes, *see* Computer codes
Code validation, 2
Computer codes, xxi
  Burgers' equation
    WENO3 code, 146
  classical hydrodynamic model
    WENO3 code, 194
  diffusion equation
    backward Euler code, 74
  list of, xxi
  2D supersonic jets
    WENO3 code, 167
  wave equation
    WENO3 codes, 172
Condition number, 110, 205
Conjugate gradients, *see* Iterative methods
Conservation laws, 71
  hyperbolic, 133, 139, 140, 159
  macroscopic, 71
  microscopic, 71
  parabolic, 73
Conservative form, 71
Conservative methods, 71
  theorem, 71, 72
  3D theorem, 72
Conserved variables, 151, 168, 195
Consistency, 9, 15
Contact wave, *see* Gas dynamics
Convergence, 9, 15
Courant number, 28, 123, 164

D

δ_{ij}, xx
$\delta(x)$, 68
Derivative approximations, 9
Derivative matrices, xx
Diffusion equation, 65, 66
  accuracy of TRBDF2, 80
  backward Euler code, 74
  boundary conditions, 65
  dynamic timestep, 80
  Fourier series solution, 69
  fundamental solution, 68
  Gaussian solutions, 59
  general solution, 69
  implicit *vs.* explicit methods, 75
  kernel, 68
  linear, 65
    TRBDF2, 67
    TRBDF2 order of accuracy, 81
  nonlinear, 66, 78, 88
    TRBDF2, 82, 83
    TRBDF2 and Newton's method, 78
  propagation speed, 69
  stability of TRBDF2, 80
Dimension by dimension splitting, 160, 164
Dirac δ function, 68
Direct solve, *see* Sparse matrix methods
Discontinuous solutions, 139
Discretization error, 13
Divided-difference formula for Δt, 42
Domain decomposition, 166
Drift-diffusion model, 188
Duhamel's principle, 62, 63
Dynamical systems, 32
Dynamic timestep, 28
  algorithm, 29

E

Elliptic PDEs
  boundary conditions, 91
  numerical methods, 91
  See also Iterative methods; Laplace's equation; Poisson's equation
Entropy, 152–154, 159, 179, 180
Entropy condition, 139, 143
ϵ_M, 12
Equation of state, 150
Equilibria
  Lorenz equations, 38
  ODE IVPs, 37
Equivalence Theorem, 9, 15, 43, 62, 104
  PDEs, 62
  proof for ODEs, 43
  proof for PDEs, 62
Euler equations, 149, 188
  classification, 179
  conserved variables, 168
  1D, 154
  2D, 167

F

Fast Fourier transform (FFT), 95
FDTD for Maxwell's equations, 124
Finite difference derivatives, 202
Finite volume methods

Index 213

multidimensional PDEs, 83
  space-time, 136, 159
First derivative, 10, 11, 202, 203
First derivative matrix, 49
Five-star stencil for the 2D Laplace/Poisson equation, 94
Floating point, 11
Flop, 13
Forward difference
  first derivative, 11, 203
Forward Euler method
  ODE IVPs, 22
  parabolic PDEs, 73
Forward time difference, 11, 203
Fourier law for heat conduction, 188
Fourier series, 70, 87, 97
  orthogonality relations, 70, 97, 203
Fourier stability analysis of PDEs, 16, 18, 75, 131, 177
Fundamental Theorem of Numerical Analysis, *see* Equivalence Theorem

G

Gas dynamics, 121, 149
  characteristic equations, 153
  conservation laws, 149
  conserved variables, 151, 168
  contact, 151, 152, 155, 166, 200
  discontinuity types, 152
  entropy advection equation, 153
  equation of state, 150
  ideal, 149
  periodic boundary conditions, 157
  rarefaction, 155, 166
  Riemann invariants, 154
  Riemann problem, 155
  Riemann problem suite, 155
  shock, 151, 152, 155, 166, 169, 195, 196, 200
  soundspeed in air, 155
  through-flow boundary conditions, 157
  wall boundary conditions, 158
Gauss-Seidel, *see* Iterative methods
GMRES, *see* Iterative methods
Godunov's method, 172
  higher-order, 174
  Reconstruct–Evolve–Average algorithm, 174
  space-time stencil, 172
Growth factor, 16
  hyperbolic PDEs
    downwind, 177
    FTCS, 132

Lax-Friedrichs, 177
Lax-Wendroff, 177
upwind, 131
ODEs
  complex, 23
  forward Euler, 22
  RK2, 45
  RK4, 45
  TR, 24
  TRBDF2, 42, 47
parabolic PDEs, 76
  backward Euler, 75
  TR, 87
  TRBDF2, 81
PDEs, 18

H

Harmonic oscillator, 27
Heat/diffusion equation, *see* Diffusion equation
Heat equation, 65
Hyperbolic conservation laws, 133
  discontinuous solutions, 139
  finite volume methods, 159
  Riemann problem, 140
Hyperbolic PDEs
  analytical tools, 121
  boundary conditions, 121, 125, 127, 157, 158
  numerical methods, 121, 122, 128, 132, 134, 135, 145, 160, 170, 172
  See also Advection equation; Burgers' equation; Gas dynamics; Wave equation
Hyperbolic *vs.* parabolic *vs.* elliptic PDEs, 57, 61, 62

I

IEEE floating point, 11
Indices
  MATLAB *vs.* C, 201
Initial value problem, 21
  numerical methods, 21, 22
Instability, 13
Inviscid Burgers equation, *see* Burgers' equation
Iteration matrix, 98
Iterative idea, 97
Iterative methods, 99
  Arnoldi iteration, 111, 112
  Chebyshev SOR, 103
  conjugate gradients, 107, 108
  conjugate gradients *vs.* steepest descent, 107

Iterative methods (*cont.*)
  conjugate gradients algorithm, 110
  Gauss-Seidel, 93, 99, 101
  GMRES, 111
  GMRES algorithm, 112
  Jacobi, 93, 99, 100
  PCG algorithm, 110
  residual, 99
  SOR, 93, 99, 101–103, 105
    ω_{opt}, 103, 105, 118
  steepest descent, 108
  SUR, 102, 103
  theory, 104
  Young's formula, 105

J
Jacobian
  finite difference, 40
  hyperbolic conservation laws, 134, 163
  nonlinear BVP, 52
  nonlinear diffusion, 81
  nonlinear Laplace equation, 114
Jacobi iterative method, *see* Iterative methods
Jets, *see* Supersonic jets

K
Korteweg-De Vries (KdV) equation, 57, 59, 63
Kronecker δ_{ij}, xx
Krylov space, 110, 111

L
Laplace's equation, 91–94
  conjugate gradients, 107
  discretization, 93
  Fourier series solution, 96
  Gauss-Seidel iteration, 101
  GMRES, 111
  iterative sweeps, 99
  Jacobi iteration, 100
  nonlinear, 113, 116
    Newton's method, 113, 116
    simulations, 116
  order of accuracy, 98
  PCG, 110
  simulations, 92, 96
  SOR
    ω_{opt}, 103
    iteration, 101
    sweeps, 103
Laplacian matrices
  memory and complexity, 95

Lax entropy condition, 139
Lax-Friedrichs (LF) method, 132, 133
  four versions, 148
Lax-Wendroff method, 134
  two-step, 135
Lax-Wendroff Theorem, 139
Layer boundary value problem, 52
Lipschitz continuous, 19
Local error, 18
Local truncation error (LTE), 18
  PDEs, 74, 128
Lorenz equations, 36
  equilibria, 38
  Jacobian, 39
L-stability, 16, 17, 22, 40, 42, 67, 73
  backward Euler, 23
  TRBDF2 method, 43

M
Machine epsilon ϵ_M, 12
MATLAB *vs.* C indices, 201
MATLAB programs, *see* Computer codes
Maxwell's equations
  FDTD, 124
Minimal degree ordering, *see* Sparse matrix methods
Mixed type PDEs, *see* Classical hydrodynamic model; Drift-diffusion model; Navier-Stokes equations
Modified midpoint rule, 45
Modified PDE, 129, 132
Multigrid, 49

N
Navier-Stokes equations, 200
  Chorin's method, 192, 193
  classification, 63
  compressible, 187
  incompressible, 188, 191
Nested dissection, *see* Sparse matrix methods
Newton's method
  backward Euler method, 39
  BVPs, 53
  nonlinear BVPs, 51
  TRBDF2 method, 40
Norms, 205
Notation, xix
Numerical codes, *see* Computer codes
Numerical errors, 13
Numerical instability, 13
N-wave, 144

Index

O
ODE BVPs, *see* Boundary value problem
ODE IVPs, *see* Initial value problem
One-step method, 21

P
Parabolic conservation laws, 73
Parabolic PDEs
    boundary conditions, 65
    numerical methods, 65, 67
    relationship to ODE IVPs, 76
    See also Diffusion equation
Parallel methods
    chaotic relaxation, 104
    domain decomposition, 166
    MPI, 166
    OpenMP, 101
PCG, *see* Iterative methods
PDEs overview, 57
Peano Existence Theorem, 32
Pendulum, 32
Physical solution, 139
Picard-Lindelöf Theorem, 32
Poisson-Boltzmann boundary value problem, 55
Poisson-Boltzmann equation, 116, 119, 191
Poisson's equation, 61, 91, 93, 94, 188, 190, 192, 194, 195
Predictor-corrector methods
    Navier-Stokes equations, 193
    ODE IVPs, 25
Principle of Uniform Boundedness, 63
Programs, *see* Computer codes

R
Rankine-Hugoniot jump conditions, 121, 137–139, 143, 144, 151, 153, 155, 166
Rarefaction wave, *see* Burgers' equation; Gas dynamics
Recommended methods, 3
    ODEs, 3
    PDEs, 5
Reconstruct–Evolve–Average algorithm, 174
Reynolds number, 192, 200
Riemann invariants, 125, 126, 154
Riemann problem, 140, 155
    Burgers' equation, 140
    gas dynamics, 155, 156, 166
Roundoff error, 13
Roundoff level, 12
Runge-Kutta methods
    ODE IVPs, 26

S
Schrödinger's equation, 57, 60–62
Second derivative, 11, 203
Second derivative matrix, 49
    eigenvalues, 86
Shaw oscillator, 34
Shock wave, *see* Burgers' equation; Gas dynamics
    loss of accuracy at, 175
SOR iterative method, *see* Iterative methods
Sparse matrix methods
    banded matrix direct solve, 49, 94, 95, 115
    minimal degree ordering, 95
    nested dissection, 95
Spectral method
    heat/diffusion equation, 70
Spectral radius, 98, 104, 105
    Gauss-Seidel iteration, 99
    Jacobi iteration, 99, 106, 117
    SOR iteration, 99, 102, 106
    SUR iteration, 102
Stability, 9, 15
    model problem for ODE IVPs, 23
    PDEs, 18
Stability of equilibria, 37
Stability region, 23
    backward Euler, 23
    forward Euler, 23
    trapezoidal rule, 23
    TRBDF2, 23
Steepest descent, *see* Iterative methods
Stiff differential equations, 40, 46, 66, 76
Strang splitting, 185
Supersonic jets
    radiative cooling, 164
    3D simulations, 164
    2D simulations, 167, 169
    WENO3 code, 167
    See also WENO3 codes; WENO3 method
SUR iterative method, *see* Iterative methods
Symbol of PDE system, 61, 63, 134, 179, 199

T
Taylor series formulas, 201
Total variation diminishing, 174
Trapezoidal rule (TR) method
    ODE IVPs, 24
    order of accuracy, 48
    parabolic PDEs, 76
TRBDF2 method
    divided-difference formula, 42
    Newton's method, 40
    ODE IVPs, 40

TRBDF2 method (*cont.*)
  parabolic PDEs, 67
  revisited for PDEs, 78
Truncation error, 13

U
Upwind difference
  first derivative, 10
Upwind method, 128
  conservative, 147
  conservative first-order, 146
  nonconservative, 146

V
Van der Pol oscillator, 30, 34
Vector norms, 205
Von Neumann stability analysis, *see* Fourier stability analysis of PDEs

W
Wave equation, 121, 124, 126
  D'Alembert solution, 127
  upwind, 58, 130
  WENO3 code, 172
  WENO3 order of accuracy, 181
WENO3 codes, 170
  Burgers' equation, 146
  classical hydrodynamic model, 194
  2D supersonic jets, 167
  wave equation, 172
WENO3 method
  grid stencil, 161
  Riemann problems, 166
  RK3 timestep, 164
  3D supersonic jets, 164
  2D supersonic jets, 167, 169
WENO method, 160
  characteristic variables, 163
  dimension by dimension splitting, 164
  Lax-Friedrichs flux splitting, 163
  positivity-preserving scheme, 164

Y
Young's formula, 105

SPRINGER NATURE

GPSR Compliance

The European Union's (EU) General Product Safety Regulation (GPSR) is a set of rules that requires consumer products to be safe and our obligations to ensure this.

If you have any concerns about our products, you can contact us on ProductSafety@springernature.com

In case Publisher is established outside the EU, the EU authorized representative is:

Springer Nature Customer Service Center GmbH
Europaplatz 3
69115 Heidelberg, Germany

The manufacturer's authorised representative in the EU is Springer Nature Customer Service Centre GmbH, Europaplatz 3, 69115 Heidelberg, Germany. If you have any concerns regarding our products, please contact ProductSafety@springernature.com

Printed and bound by CPI Group (UK) Ltd, Croydon, CR0 4YY

25/03/2026

02078171-0006